GO BIG BY GETTING SMALL

The Application of Operational Art By Special Operations In Phase Zero

Brian S. Petit

Foreword by Admiral (US Navy, Ret.) Eric Olson

The opinions expressed in this manuscript are solely the opinions of the author and do not represent the opinions or thoughts of the publisher. The author has represented and warranted full ownership and/or legal right to publish all the materials in this book.

Going Big by Getting Small: The Application of
Operational Art By Special Operations In Phase Zero
All Rights Reserved.
Copyright © 2013 Brian S. Petit
v4.0

Cover Photo © 2013 Brian Petit. All rights reserved - used with permission.

This book may not be reproduced, transmitted, or stored in whole or in part by any means, including graphic, electronic, or mechanical without the express written consent of the publisher except in the case of brief quotations embodied in critical articles and reviews.

Outskirts Press, Inc.
http://www.outskirtspress.com

ISBN: 978-1-4787-0385-3

Library of Congress Control Number: 2013911819

Outskirts Press and the "OP" logo are trademarks belonging to Outskirts Press, Inc.

PRINTED IN THE UNITED STATES OF AMERICA

This book is dedicated to quiet professionals everywhere, military and civilian alike, whose seemingly small, yet finely calibrated words and actions do more to protect our nation than we will ever know.

Table of Contents

1 INTRODUCTION 1
   Special Operations Phase Zero.......................................6
   Research Question 8
   Research Methodology 10
   Audience 11
   Summary of Results 12
   Study Organization..15

2 BACKGROUND...17
   Section I. Understanding Operational Art 17
      Defining Operational Art...17
      A Brief History of Operational Art............................19
      The Modern Need for Operational Art......................24
      Designing Operational Art: Doctrinal Methods.........26
      Operational Art: Modern Demands 31
   Section II. Understanding Special Operations..................32
      Defining Special Operations 32
      SOF Command Structure..34
      Historical Development of Special Operations 36
      Special Operations Theory and Doctrine 39
      Special Operations: Contemporary Application 47
      Summary 52
   Section III. Understanding Phase Zero 53
      Definition of Phase Zero 53
      Historical Development of Phase Zero.......................57

  The Development of Diplomacy 61
  Theories of Diplomacy 63
  Phase Zero: Contemporary Application 67
  The Department of Defense Phase Zero Framework 68
  Summary 72

3 PHASE ZERO TENSIONS 73
  Section I. The Sources of Phase Zero Tensions 74
  Section II. Synchronization Frameworks 79
  Section III. Synchronization Challenge One: Policy 84
   Incremental Policy Vignette: Yemen 85
   Political Policy Vignette: Special Operations
    Engagement in Thailand 98
   Legislative Policy Vignette:
    1997 Leahy Amendment 91
   Summary: Policy and Engagement 105
  Section IV. Synchronization Challenge Two:
   Programs 107
  Section V. Synchronization Challenge Three: Posture .. 113
  Conclusion: Phase Zero Tensions 119

4 CASE STUDY 120
  Background: Colombia 122
  US Southern Command and US Policy 125
  US Special Operations Forces Engagement 127
  US Hostage Crisis and SOCSOUTH Assessment 130
  SOCSOUTH Operational Approach 132
  Summary 136

5 ANALYSIS OF SPECIAL OPERATIONS
  PHASE ZERO OPERATIONAL ART 137
  Termination 139
  Military End-State 141

	Center of Gravity	144
	Decisive Points	148
	Direct and Indirect Approach	151
	Anticipation	154
	Operational Reach	157
	Culmination	158
	Arranging Operations	160
	Summary	164
6	CONCLUSIONS	166
	Areas for Further Study and Recommendations	174
	Implications	176
	LIST OF ABBREVIATIONS	178
	BIBLIOGRAPHY	182

List of Figures

Figure 1.	Arranging Chain	14
Figure 2.	Special Operations Expressions of Joint Elements of Operational Design	15
Figure 3.	Elements of Operational Design, Joint Publication 5-0, III-18 Figure III-9	22
Figure 4.	Operational Art, Joint Publication 5-0, III-2 Figure III-1	28
Figure 5.	Notional Operation Plan Phases, Joint Publication 5-0, III-39 Figure III-16	55
Figure 6.	Depiction of Power Behavior Spectrum, adapted from Joseph S. Nye, Jr., *The Future of Power*, 21	66
Figure 7.	Effective Phase Zero Outcomes	80
Figure 8.	Map of Yemen, CRS, July 2012	86
Figure 9.	Map of Indonesia, CRS, November 2011	93
Figure 10.	Map of Celebes Sea Region, USGS 2011	95
Figure 11.	Map of the Kingdom of Thailand, CRS, June 2012	99
Figure 12.	Department of Defense Security Assistance Programs	108
Figure 13.	Map of Colombia, CRS, November 2012	121

Figure 14.   Special Operations Expressions of Joint
             Elements of Operational Design.....................138
Figure 15.   Arranging Chain 163

List of Photos

Photo 1 34
   A US special operations soldier combat advises friendly nation forces in Africa (US Army photo, authorized use).

Photo 2 36
   Special operations employ advanced medical capabilities to increase host nation medical capacity and, when prudent, deliver treatment directly to partners and populations (US Army photo, authorized use).

Photo 3 46
   Special operations Cultural Support Teams (CSTs) are small, female-led teams with specialized skills to engage local populations (US Army photo, authorized use).

Photo 4 49
   Special operations combat advisors conferring with partnered force soldiers (US Army Photo, authorized use).

Photo 5 ....
   A special operations soldier patrols with an Afghan local policement (US Army photo, authorized use).

Photo 6 79
   Civil affairs teams apply unique capabilities to assist remote populations with livelihood and health issues (US Army photo, authorized use).

Photo 7 102
   A US Army Special Forces soldier conducts small unit tactics training with partnered forces (US Army photo, authorized use).
Photo 8 112
   JCET programs are the main vehicle for USSOF Phase Zero engagements with friendly nations (US Army photo, authorized use).
Photo 9 118
   Skilled partnered force training exchanges employ methods suitable and sustainable for partnered nation forces (US Army photo, authorized use).
Photo 10 130
   Many Phase Zero engagements include small unit tactics (US Army photo, authorized use)

Acknowledgements

I am grateful for the opportunity, in my 22nd year of military service, to spend an academic year studying, thinking, and writing. Thank you to the US Army for this unique educational opportunity. This book was written while I was a student in the Advanced Operational Art Studies Fellowship (AOASF) at the School of Advanced Military Studies (SAMS) at Fort Leavenworth, Kansas. The AOASF Director, Dr. Pete Schifferle, deserves recognition for building and leading the curriculum that helped me synthesize many of the ideas contained in this study.

I owe a great deal of thanks to those who freely contributed their ideas to this study, either directly or indirectly. Those with significant influence on this book are Eric Walker, Mike Rogan, Jim Tennant, Dave Maxwell, Mark Rosengard, Bob Jones, Josh Walker, Joe Mouer, Bill Moore, Ken Gleiman, Kalev Sepp, and Greg Wilson. Thank you to Joe Celeski, who took a particular interest in this project and generously offered his insights into the larger, strategic questions regarding special operations. A personal thanks to Mike Kenny for the rich daily dialogue that shaped much of this work.

The final and most important thank you goes to my best friend, editor, critic, mentor, and wife: Dr. Jodi Breckenridge Petit. Without Jodi's direction and guidance, this project would have never reached completion.

Given these acknowledgements, any errors in this work are mine alone.

Foreword

Much of the literature on special operations is dominated by headline-making missions: deep raids, harrowing firefights, close combat actions, and similar acts of singular daring. These are often fascinating and powerful stories, and the public appetite for them seems limitless. Yet there is another narrative on special operations, told less often and with greater difficulty, but arguably more central to the long-term interests of our great nation and stability around the globe. In *Going Big By Getting Small: The Application of Operational Art by Special Operations in Phase Zero*, Brian Petit makes a seminal contribution to this less-public side of US special operations and the forces that conduct them.

This book superbly examines the key intersection of strategy, policy, diplomacy, and special operations. Colonel Brian Petit, a broadly experienced US Army Special Forces officer whom I first met when he was leading an operational unit in a very complex combat environment, draws on history, theory, and doctrine to make a convincing case for a special operations-centric operational art. His book is essential reading for those who seek to understand how the US can wisely achieve its objectives abroad with a joint-force enabled special operations capability. This is not a story of raids and firefights. It is a narrative on the artful link of tactics to strategy, compelled in part by special operations practitioners who, while highly skilled in the conduct of war, are also most able to prevent it.

This in-depth look at steady-state US engagement abroad before the shooting starts is of vital importance. The current US strategic quandary is how to sustain or increase our standing in the world as power paradigms shift under our feet. Common to all analyses on this topic is one consistent theme: limit large wars and optimize pre-conflict engagements abroad. Brian Petit shows exactly how the strategy and tactics of engagements in non-wartime environments can better realize US strategic aspirations. *Going Big by Getting Small* is a rare and timely work. By capturing the art of the possible in the realm of lightly footprinted engagements, greater attention is given to the strategic utility of US special operations. Daring raids and rescues notwithstanding, this conversation is long overdue.

Admiral (US Navy, Ret.) Eric Olson
Former Commander,
United States Special Operations Command
May 2013

1
Introduction

Grand strategy is the singular, defining idea that orients a nation's power and influence abroad. The process of grand strategy involves marshaling and integrating national resources in pursuit of a nation's interests, security, and future prosperity.[1] Equal parts ideology and policy, national grand strategies cannot be ordered into existence by executive fiat. Some take flight and some do not. Past American grand strategies include expansionism, isolationism, Rooseveltian imperialism, Wilsonian idealism, and Cold War containment.[2] Twenty years after the Cold War, the United States (US) lacks a grand strategy.[3] This absence reflects the difficulty

---
1 Edward Luttwak, *The Logic of War and Peace* (Cambridge, MA: Belknap Press, 2003), 207-217. Henry Mintzberg and James Brian Quinn, Readings in the Strategy Process, 3d ed. (Upper Saddle River, New Jersey: Prentice Hall, Inc.), 367-78.
Everett Carl Dolman, Pure Strategy: Power and principle in the space and information age, (New York: Frank Cass, 2005), 26-30.
2 Walter A. McDougall, "The Constitutional History of U.S. Foreign Policy: 222 Years of Tension in the Twilight Zone," Center for the Study of America and the West at the Foreign Policy Research Institute (based on lectures delivered at the Annenberg Summer Teacher Institute, National Constitution Center, Philadelphia, PA, July 27, 2010), September 2010.
3 Rosa Brooks, "Obama Needs a Grand Strategy," Foreign Policy, January 23, 2012. http://www.foreignpolicy.com/articles/2012/01/23/obama_needs_a_grand_strategy (accessed July 23, 2012).

in defining an overarching idea for the US in today's complex, multipolar, and power-diffused world. Lacking a grand strategy, the US executes multiple *strategies* that compete in a Darwinian forum to gain prominence, influence policies, and determine resource commitment. Of these competing strategies, one means has emerged as a poor substitute, yet a reliable surrogate, for grand strategy: engagement.[4]

Engagement is broadly defined as "the active participation of the United States in relationships beyond our borders."[5] Engagement is conducted by US diplomatic, defense, or development agencies to promote relationships, programs, and progress deemed mutually beneficial to both the US and its friends and allies. For adversaries and competitors, engagement displays transparency and dialogue with strategic signaling that demonstrates US power, intent, and capability without unnecessary provocations.[6]

Peacetime engagement events represent US policy-in-action. Engagement events, enacted in the diplomatic, military, information, and economic realms, comprise the ways and means to achieve strategic ends.[7] In military parlance, engagement occurs in "Phase Zero" or the pre-crisis environment in which state relations are peaceful and routine.[8] Though all

---

4   The White House, *National Security Strategy of the United States of America, May 2010* (Washington DC: The White House, 2010). http://www.whitehouse.gov/sites/default/files/rss_viewer/national_security_strategy.pdf (accessed September 26, 2012)
5   National Security Strategy, 2010, 19.
6   Richard L. Kugler, "New directions in national security strategy, defense plans, and diplomacy: a review of official strategic documents" (Washington DC: National Defense University Press), 2011, 2-5.
7   The term "engagement events" includes nearly all US actions abroad in peacetime environments. For the US Department of State, a key engagement event is a visit by the Secretary of State, which often is accompanied by a US resource commitment through new or existing programs. Different forms of engagements are discussed, in detail, in chapter three.
8   Department of Defense, Joint Publication (JP) 5-0, Joint Operation Planning, (Washington DC: Government Printing Office, 2011), III-42. Joint doctrine employs a five-phase model for planning joint operations. Each phase in the five-phase joint phasing model by title (e.g, Shape or Deter) with the phase number in parenthesis (e.g.,

military engagements can inherently be associated with preparation for warfare, the explicit purpose of many Phase Zero military engagements is to *prevent* war.[9] The strategic logic is that where engagements exist, the US is *in dialogue*, applying a spectrum of efforts – short of warfare - to shape the strategic environment to its advantage.

Engagement is arguably the strongest enduring idea of the past twenty years of National Security Strategies (NSS) and foreign policy directives.[10] The 1995 NSS of "Engagement and Enlargement" codified US peacetime engagement as the leading line of effort to sustain US security. The 2010 NSS strikes a similar tone, citing engagement 43 times.[11] Engagement remains *the* central idea for securing our nation – an aspirant toward grand strategy. Despite this emphasis, engagement

---

Shape (Phase 0) or Deter (Phase I)). This study uses the semantic distinction of "Phase Zero" instead of "Phase 0." This distinction, common in the literature of Phase Zero discussions, is used as a descriptor of an environment and a condition (short of war) as opposed to simply a linear, numbered phase as part of a five-phase model.

9   The annual posture statements of the US geographical (and functional) combatant commands are the most useful unclassified sources that articulate the totality and purpose of US military engagement programs. The posture statements are presented to Congress annually by the commanding general of the combatant command and are easily accessible online by searching "combatant command posture statement."

10 The White House, *National Security Strategy of the United States, January 1993,* (Washington DC: The White House, 1993). The 1993 NSS defines "engagement" as the overarching idea of national security in the post Soviet Union world.
The White House, A National Security Strategy of Engagement and Enlargement, February 1995.
The White House, A National Security Strategy for a New Century, October 1998.
The White House, The National Security Strategy of the United States, 2002.
The White House, The National Security Strategy of the United States of America, 2006.
The White House, National Security Strategy, 2010.
Following the September 11, 2001 attacks on the United States by Al Qaeda, engagement remained a feature of national security strategies but arguably became subordinate in emphasis to counterterrorism. The 2010 NSS returned, in spirit and tone, to the 1993 NSS by elevating engagement as the main idea underpinning the goals and aspirations of the US.

11 Gleiman, LTC Jan Kenneth. "Operational Art and the Clash of Organizational Cultures: Postmortem on Special Operations as a Seventh Warfighting Function." Monograph, School of Advanced Military Studies, United States Army Command and General Staff College, Fort Leavenworth, KS, 2011, 1. Gleiman cites the use of the term "engagement" 43 times in the 2010 NSS.

has failed to coalesce into an idea or a method from which a grand strategy can grow.

One reason for this failure is the difficulty of arranging engagements in time, space, and purpose to construct or support a cohesive strategy.[12] The US armed forces, the executor of the military aspect of strategy, are often deficient in employing engagements to achieve strategic ends.[13] Despite the clear US strategic reliance on all forms of engagement, the crafting of military engagements to achieve strategic objectives is a practice with an insufficiently coherent *operational art*, the skillful linking of tactics to strategy. Without a clear operational art concept, the military component of engagement is an unreliable method to accomplish strategic objectives. For military engagement to fulfill the aims of US security strategy and propel "engagement" into an aspirant of grand strategy, an improved concept and practice of Phase Zero operational art is needed.

The military conception of operational art remains principally in the domain of high intensity, combined arms warfare: the harmonious synchronization of military power to destroy an opposing military.[14] This notion of operational art is valid for force-on-force warfare but is less relevant for an engagement-centric strategy. For the US to manage our threats and realize our aspirations abroad, we need a

---

12 Strategists such as Shimon Naveh and Everett Carl Dolman discuss the difficulty of forming strategy inside of complex, adaptive systems like global security structures. Engagements, which are incremental and cumulative rather than decisive, represent one of many independent variables acting to influence the overall system.

13 General Rupert Smith, *The Utility of Force: The Art of War in the Modern World*, (New York: First Vintage Books, 2008), 308-334. Smith's book is a polemic about the inadequacy of today's modern military apparatus to contend with nature of modern warfare, or "war amongst the people."

14 Michael D. Krause and R. Cody Phillips, ed. *Historical Perspectives of the Operational Art*, (Washington DC: Center for Military History, 2005). In this anthology, thematically "the operational artist's center of gravity is the mass of the enemy military's force and its ability to command and control its forces," v.

INTRODUCTION

complementary operational art better suited for Phase Zero environments.

An operational art for engagement is forming, if slowly. The contours of engagement-centric operational art are emerging in form and practice across the armed services, in all combatant commands (COCOMs), and within the interagency. Among the leaders in crafting effective engagement campaigns are US Special Operations Forces (USSOF). Building on its core competency of foreign military engagement, USSOF are engineering improved form and function in the middle ground between tactics and strategy.[15] USSOF – led by seven geographic Theater Special Operations Commands (TSOCs) - are designing and executing unique operational approaches in Phase Zero.[16] USSOF Phase Zero campaigns, conducted under combatant command and ambassadorial authorities, possess a distinct, if difficult to define, character.[17] USSOF Phase Zero campaigns are not traditional operational-level military campaigns, yet they are significantly more than the logical linkage of small tactical actions. This operational art, however nascent, is gradually finding its form. Defining the *how* and *what* of this putative special operations operational art is the topic of this study. The author's aim is to discover what aspects of operational art are required for Phase Zero, with a particular focus on special operations.

---

15 John Andreas Olsen and Martin Van Creveld, ed.,*The Evolution of Operational Art*, (Oxford: Oxford University Press, 2011). Olsen and Van Creveld classify the operational art as the "gray area between tactics and strategy," introduction, ix.

16 The seven Theater Special Operations Commands (TSOCs) are discussed in chapter two.

17 Department of Defense, Joint Publication (JP) 5-0, *Joint Operational Planning*, (Washington DC: Government Printing Office, 11 August 2011, III-42. Campaigns are defined as a "series of related major operations aimed at achieving strategic and operational objectives within a given time and space." GL-6.

Special Operations Phase Zero

USSOF currently conducts engagements in over 77 countries.[18] Nearly all these USSOF engagements occur in the Phase Zero environment declared as routine military activities, informally labeled "upstream engagement."[19] In these environments, special operations are less a tool for war than a method of statecraft for achieving a favorable international order. The aim of special operations in Phase Zero is to offer combatant commanders, ambassadors, and host nations the right instrument to meet their security, diplomatic, or political challenges. As the strategic setting for the US shifts from large-scale expeditionary wars to multilateral cooperation, the skilled use of USSOF in upstream engagements holds great potential. Realizing this potential requires improved precision, nuance, and wisdom from civilian and military leaders arranging and directing special operations missions.[20] To further aid in shared understanding, a descriptor called "special operations power" is emerging in theory and practice. This study adopts the definition of special operations power proposed in 2013 by Colonel (USA Ret.) Joseph D. Celeski.

---

18 Admiral William H. McRaven, Commander, US Special Operations Command, address to Sovereign Challenge IX, San Jose, CA, June 6, 2012. The author was present at ADM McRaven's address. McRaven cited the presence of USSOF in 77 countries. However, not all of these SOF engagements are in Phase Zero environments. The significant USSOF presence in Afghanistan, for example, cannot be considered a Phase Zero environment.

19 The phrase "upstream engagement" conveys the idea of assessing and addressing problems in their infancy "upstream" in order to preclude having to contend later with major problems, up to and including costly combat operations. The British informally use the term "upstream engagement;" it has no formal doctrinal definition in US or British doctrine at this time.

20 Admiral William H. McRaven, "Posture Statement, Admiral William H. McRaven, USN, Commander, United States Special Operations Command, before the 112th Congress, United States Senate, March 06, 2012." http://www.socom.mil/Documents/2012_SOCOM_POSTURE_STATEMENT.pdf (accessed September 23, 2012).

The physical force (SOF) + a theory (or theories) for its use in some medium of warfare (or domain, or environment, with focus on the Human Domain) + agreed upon way of how to apply it through codifying doctrine and principles and procedures for employment derived from experience and its own unique objectives (*Special Warfare*) + a guaranteed way to influence, coerce, compel, impose will on competitors (special operations missions and tasks – how to apply these abilities and functions to achieve effects); then minus limitations and restrictions on the use of special operations and SOF = **Special Operations Power**[21]

The special operations power paradigm used in Phase Zero is evolving under the confluence of four modern-day phenomena: a national strategic emphasis on partner engagement and multilateralism, a globally oriented and organizationally maturing United States Special Operations Command (USSOCOM), the proliferation of non-state threats less susceptible to conventional warfare, and the global diffusion of power.[22] Given the dynamic convergence of these factors, the search for how special operations sees operational art may reveal compelling insights on the broader ideas of US national strategies abroad. In turn, these revelations may stimulate new thinking on US grand strategy.

---

[21] COL (USA Ret.) Joseph D. Celeski, LTC (USA Ret.) Timothy S. Slemp, COL (USAF, Ret.) John D. Jogerst, *An Introduction to Special Operations Power: Origins, Concept, and Application*, draft monograph, April 2013, 2. Definition used by permission of the authors.

[22] Joseph S. Nye, Jr., *The Future of Power*, (New York: PublicAffairs, 2011). Nye uses the phrase "diffusion of power" to describe the shifting of power in the 21st century from singular, hegemonic nation-states with powerful militaries to an environment where a multitude of nations, actors, interest groups and populations have access to levers of power previously controlled by powerful nations.

Research Question

This study aims to answer the question: What are the special operations elements of operational art in Phase Zero? To answer this question, three supporting questions are examined. How do the synchronized application of special operations operations, actions, and activities in Phase Zero constitute a military campaign? What are the unique characteristics that distinguish special operations in Phase Zero campaigns from traditional military campaigns? How do special operations in Phase Zero campaigns support the broader theater campaigns?

A critical assertion prefacing this study is that there are distinct and unique applications of special operations power at the operational art level, in Phase Zero, which are under-explored in current military campaign design. This assertion is supported by three compelling developments: special operations global reach, a shifting strategic security environment, and the revised US military contemplation of the land and human domains.[23]

First, the systemic and persistent global employment of special operations in undeclared hostile regions in over 75 countries is evidence enough to explore special operations in Phase Zero as "campaign-like" within operational art. Beyond

---

23   Major General Bennet S. Sacolick and Brigadier General Wayne W. Grigsby, Jr., "Special Operations/Conventional Force Interdependence: A Critical Role in 'Prevent, Shape, Win,'" June 2012, *Army Magazine*, Volume 62, No. 6, 39-42. http://www.ausa.org/publications/armymagazine/archive/2012/06/Pages/default.aspx (accessed October 03, 2012). USASOC introduced the concept of the 'human domain' in the 2012 initial draft of *Army Doctrinal Publication 3-05, Army Special Operations*. On May 15, 2012, the Doctrine 2015 General Officer Review Board delayed the inclusion of the term in Doctrine 2015 publications to allow for the further development and consideration of the idea. This article defines the human domain as "The totality of the physical, cultural, and social environments that influence human behavior to the extent that the success of any military operation or campaign depends on the application of unique capabilities that are designed to win in population-centric conflicts," 40.

special operations, all six geographic combatant commands are equally in pursuit of improved Phase Zero operational art paradigms.[24] A better articulation of the special operations contribution, logic, and methods will assist both SOF and the COCOMs in optimizing the use of engagements in accomplishing strategic objectives.

Second, the US strategic security context needs amended concepts and methods to adapt to a shifting threat environment. Professor of military strategy Max Manwaring classifies US adversaries as increasingly "rhizomatic" with an "apparently hierarchical system above ground- visible in the operational and political arenas, and with another system centered in the roots underground."[25] Further shifts in this strategic environment include hybrid adversaries, ubiquitous communications tools, global economic malaise, reduced NATO military capability, violent social identity conflicts, a contested cyber space, and stretched US military capacity.[26] Traditional US military power projection methods and major theater war are poor options to address many of theses challenges.

Finally, in 2012, US forces that principally operate on land – The US Army, The United States Marine Corps, and Special Operations Forces - combined to create the office of Strategic Landpower.[27] The publicized intent is to shape both current

---

[24] The author visited five geographic combatant commands and USSOCOM, occurring between October 2012 and February 2013.

[25] Max Manwaring, "Ambassador Stephen Krasner's Orienting Principle for Foreign Policy (and Military Management) – Responsible Sovereignty," (Carlisle Barracks, PA: US Army War College) Monograph, Strategic Studies Institute, April 2012, 56. http://www.strategicstudiesinstitute.army.mil/pubs/people.cfm?authorID=18 (accessed September 28, 2012).

[26] Max Manwaring, "The Strategic Logic of the Contemporary Security Dilemma," (Carlisle Barracks, PA: US Army War College), Strategic Studies Institute, December 01, 2011, 58. http://www.strategicstudiesinstitute.army.mil/pubs/display.cfm?pubID=1091 (accessed September 13, 2012).

[27] Steven Metz, "Strategic Horizons: U.S. Army Prepares for the Human Domain of War," November 7, 2012, World Politics Review. http://www.worldpoliticsreview.com/articles/12481/strategic-horizons-u-s-army-prepares-for-human-domain-of-war

operational constructs and future force structure thinking and contend with the emerging concept of the "human domain."[28] Defense analyst Steven Metz stated "the Office of Strategic Landpower will attempt to integrate the cross-cultural psychological skills of Special Operations Forces into the military's land forces writ large."[29] With focused, joint emphasis on strategic landpower constructs, a refined operational art concept for special operations Phase Zero campaigns can aid the future coupling of US Army, Marine Corps, and SOF capabilities in future upstream engagements.

Research Methodology

Operational art for special operations in Phase Zero is explored through three vignettes and one case study. The vignettes examine the intersection of policy and special operations in Yemen, Indonesia, and Thailand. The case study examines the Special Operational Command South (SOCSOUTH) campaign in Colombia (1998-2008), culminating in the 2008 Colombian SOF hostage rescue of American citizens. The vignettes and case study are supplemented by personal interviews. Combined, these perspectives help synthesize the intersecting ideas-in-practice of operational art, Phase Zero, and special operations. Once synthesized, acknowledging the differences inherent in a special operations application of operational art aids in ensuring the success of special operations at the geo-strategic, strategic, and operational art levels.

The thirteen elements of operational design in Joint Publication 5-0, *Joint Operation Planning*, are used as evaluation

---

(accessed January 13, 2013).
28 Sacolick and Grigsby, 40, and Metz, 2-3.
29 Metz, 3.

criteria to analyze the Colombian case study. The analysis method is then reversed: the Colombian case study is used to evaluate the thirteen elements of operational design.[30] This method first reveals the logic and design of the US-advised Colombian operational art construct. It then examines the thirteen elements of operational design as a suitable design template for conceptualizing and building a special operations Phase Zero-like campaign. The first goal is to identify and understand the special operations expression of each element of operational design. Secondly, the examination of the thirteen elements of operational design are analyzed and, if appropriate, modified in order to improve the doctrinal template for conceptualizing how operational art is used by USSOF in Phase Zero. The result is a recommendation to modify the elements of (US joint doctrine) operational design that better captures the logic, design, and articulation of special operations expressions of Phase Zero operational art.

Audience

This study is directed at two audiences. The first audience is the USSOF community. The strategic context demands that USSOF maximize their utility in support of our nation's strategies, wars, conflicts, and peace. To do so, USSOF must further develop their professional expertise in operating in non-wartime environments under the strategic aegis of a US ambassador. A large portion of SOF is tactically designed for this very purpose. The challenge is translating SOF tactical

---

30   Department of Defense, Joint Publication (JP) 5-0, *Joint Operation Planning*, (Washington DC: Government Printing Office, 2011), III-18. The elements of operational design are listed in chapter two, figure 1. The elements are termination, end state, objectives, effects, centers of gravity, decisive points, lines of operations and lines of effort, direct and indirect approach, anticipation, operational reach, culmination, arranging operations, forces and functions.

excellence into an appropriate and effective operational form, within a combatant command construct, that achieves US strategic objectives.

The second audience is strategists, academics, joint force leaders, and interagency partners. This helps the audience in understanding USSOF Phase Zero contributions as an instrument of power for the purposes of achieving US political goals abroad. Secretary of State Hillary Rodham Clinton stated "We need Special Operations Forces who are as comfortable drinking tea with leaders as raiding a terrorist compound. We also need diplomats and development experts who understand modern warfare and are up to the job of being your partners."[31] A nuanced understanding of special operations is paramount to ensure that civilian and military leaders from outside of the SOF community contribute informed opinions, advice, and decisions where special operations are considered.

To achieve greater understanding for both audiences, the literature on special operations must extend beyond the SOF-as-commando narratives and show the links between USSOF and diplomacy, security, and statecraft. By addressing the literature gap of operational art for special operations in Phase Zero, this study attempts to link SOF engagements with US strategic aspirations.

Summary of Results

Special operations Phase Zero campaigns are a nuanced application of specialized forces, often in domain gaps, that

---

31   Speech by Secretary of State, Hillary Rodham Clinton, May 23, 2012, US Department of State website, www.state.gov/secretary/rm/2012/05/190805.htm (accessed on December 15, 2012).

INTRODUCTION

increase US operational reach and influence in environments where diplomacy is the leading art. The evidence suggests that USSOF is crafting new campaign modes guided by revised operational art constructs. These campaigns, simultaneously developing in different worldwide regions, blend consistently applied special operations power with improvisational, localized approaches. This finding is supported by six conclusions.

- special operations execute distinct Phase Zero campaigns propelled by an innovative application of operational art
- the US Department of Defense joint doctrine elements of operational design require modifications to better guide USSOF Phase Zero campaign planning
- USSOF campaigns contain logic that appears paradoxical to the generally applied principles of the use of US military force
- Phase Zero presents a particularly difficult environment to formulate and apply an operational art
- special operations Phase Zero operational art combines supply chain management-style structure with network logic; the hybrid is a type of "arranging chain" that makes operational artistry possible over vast time, distance, cultural, and programmatic spans (Figure 1)
- despite their clear distinctions, USSOF campaigns are inextricably tied to and in support of the macro strategy of the combatant commander and the US country team.

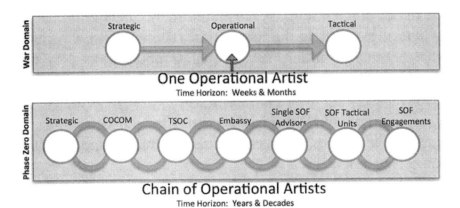

Figure 1. Arranging Chain

Operational art for special operations in Phase Zero constitutes a blend of art and science that is executed with non-traditional command arrangements, various authorities, small teams, extended timelines, and often, paradoxical logic. The SOF expressions of operational art in Phase Zero environments show a clear lineage to classic operational art, yet they possess substantive distinctions in logic, method, implementation, and effect. This study recommends modifications to nine of the thirteen US joint doctrine elements of operational design (Figure 2). These modifications are recommended as replacements for, modifications to, or expressions of their joint doctrine counterpart terms. Adjusting doctrine for operational art Phase Zero thinking has a two-fold purpose. First, to provide better guideposts for the crafting of special operations approaches appropriate in environments, short of war, where host nation sovereignty is paramount. Second, to juxtapose special operations logic with widely-used joint doctrine principles to clearly identify the commonalities and differences. Revised thinking about operational art in Phase Zero aims to achieve improved strategic utility from engagements. Within the grand scheme of US foreign engagements, special operations demonstrate a range of options for consideration. These options go beyond the well-publicized singular acts of military daring. Special operations options, both scalable and differentiated, include the conceptualization and implementaton of full campaigns.

Joint Doctrine Elements	SO Phase Zero Expressions
1. Termination	Transition
2. Military End-State	Position of Continuing Advantage
3. Objectives	Objectives
4. Effects	Effects
5. Center of Gravity	Right Partner, Place, Time (R3)
6. Decisive Points	Decisive Relationships
7. Lines of Op/Effort	Lines of Op/Effort
8. Direct & Indirect Approach	Special Warfare & Surgical Strike
9. Anticipation	Assessments & Programs
10. Operational Reach	Access & Location
11. Culmination	Saturation
12. Arranging Operations	Arranging Chain
13. Forces & Functions	Forces & Functions

Figure 2. Special Operations Expressions of Joint Elements of Operational Design

Study Organization

Chapter one introduces the topic, presents the research question and thesis, and summarizes the research design and methodology. The second chapter defines and discusses the three main components of this study: operational art, special operations, and Phase Zero. Chapter three surveys three vignettes that illustrate the tensions between US policy and USSOF Phase Zero actions. The vignettes offer insight on the environment in which unique applications of Phase Zero operational art attempt to take form. Chapter four examines the singular case study of the USSOF Phase Zero campaign in Colombia. Chapter five analyzes the USSOF expression of operational art in the case study and evaluates the research outcomes. Chapter six concludes with implications and recommendations.

2
Background

Chapter two defines and discusses the three interlocking topics examined in this study: operational art, special operations, and Phase Zero. Each topic is examined individually through the lens of history, theory, doctrine, and modern practice. The analysis of each topic in this chapter precedes the synthesis and evaluations made in chapters three through six.

Section I. Understanding Operational Art

Defining Operational Art

Current US joint doctrine defines operational art as "the creative thinking used to design strategies, campaigns, and major operations and to organize and employ military force."[32] In the simplest terms, operational art is the purposeful linkage of tactics to strategy. Strategist Colin Gray

---

32   JP 5-0, I-5.

states "strategy is the bridge that connects politics to military power."[33] Extending this same logic, operational art is the bridge that meaningfully connects the use of engagements, battles, and campaigns to accomplish strategic objectives.[34]

The inception of military campaigns conducted by industrial scale armies with vast geographic, logistic, and control challenges introduced an intermediate level of complexity between tactics and strategy.[35] Thus, operational art was born out of the necessity to arrange training, tactics, logistics, operations, and campaigns on a macro scale in pursuit of strategic ends.[36] In essence, operational *art* grew from the emergence of the operational *level* of war.[37] Over time, operational art has gained acceptance as a practice that is not defined by a specific level of war, but rather as the cognitive and practical connection between campaigns, operations, and actions at all levels to the attainment of strategic goals.[38]

---

33   Colin S. Gray, *Strategy For Chaos: Revolutions in Military Affairs and the The Evidence of History*, (Portland: Frank Cass Publishers, 2002), 92.

34   The term "engagements" has two meanings: the first is a "tactical conflict, usually between opposing lower echelon maneuver forces" (Joint Publication 1-02); the second meaning, not normally associated with battlefield military tactics but germane to this study, is the engagement defined in the introduction of this paper ("the active participation of the United States in relationships beyond our borders").

35   Martin van Creveld, "Napoleon and the Dawn of Operational Warfare," in *The Evolution of Operational Art*, ed. John Andreas Olsen and Martin van Creveld, (New York: Oxford University Press, 2011), 9-34.

36   van Creveld, 1-3.
     Bruce W. Menning, "Operational Art's Origins," in *Historical Perspectives of the Operational Art*, ed. Michael D. Krause and R. Cody Phillips, (Washington DC: Center for Military History, 2005), 4-18.

37   Menning, 4-6.
     Alexsandr Svechin, *Strategiia*, 1927. Translated, edited and published as *Strategy* by Kent D. Lee, (Minneapolis: Eastview Publications, 1992), 5. Some credit Svechin as the originator of the term and concept of "operational art" as associated with an intermediate level of war existing between strategy and tactics.

38   Department of the Army, Army Doctrinal Publication (ADP) 3-0 *Operations* (Washington DC: US Government Printing Office, October 2011). US Army doctrine now states that operational art can occur at lower tactical levels, such as the battalion. "Operational art is not associated with a specific echelon or formation, nor is it exclusive to joint force commanders. Instead, it must apply to any formation that must effectively arrange multiple, tactical actions in time, space and purpose, to achieve a strategic objective, in whole or in part," 9.

BACKGROUND

A Brief History of Operational Art

The massive growth of armies in the Napoleonic era of warfare is itself a compelling reason to consider early 19th century warfare the dawn of the operational art era.[39] Few dispute that Napoleon Bonaparte (1769-1821) is among the first and the best of history's operational artists.[40] If Napoleon is not the first true operational artist, then his relation to operational art is equivalent to that of George Washington to the US Presidency: an indelible imprint of genius, leadership, vision, and precedent that both defined the idea and set the measurement.

Napoleon's *Grand Armee* introduced a scale never before witnessed in warfare. The 1812 French campaign into Russia committed 600,000 French and allied troops on a transcontinental march *as part of* a French imperial strategy. Outcomes in Russia notwithstanding, Napoleon's mastery in mass-scale warfare required innovation in all areas: recruiting, organization, combined arms effects, maneuver, sustainment, command and control, the advent of the corps, and the use of battles and campaign to achieve strategic objectives.[41]

How is Napoleonic operational art relevant to a discussion on modern day special operations? Napoleon's battlefield tactics may well be an outdated field of study, but his conduct of European land warfare from 1799 to 1815 provided the

---

39 van Creveld, 17-19.
40 The Duke of Wellington, who defeated Napoleon at Waterloo in 1815, is among those who might dispute Napoleon's rank as a top operational artist. Wellington is reported to have labeled Napoleon as a "butcher." Author visit and tour to Waterloo, Belgium, October 2012.
41 Carl Von Clausewitz, *Vom Kriege*, 1832. Translated, edited and published as *On War* by Michael Howard and Peter Paret (Princeton: Princeton University Press 1984). *On War* is replete with references, analysis, praise and criticism for Napoleonic campaigns. Much of the tactical discussions, now considered outdated, involve the French application of the corps, logistics, reserves, command and control and other tactical operating systems and concepts.

principle subject of study for the enduring theories of Baron Antoine de Jomini (1779-1869) and Carl Von Clausewitz (1780-1831).[42] Jominian and Clausewitzean theories remain the intellectual and doctrinal underpinning of modern US military theory and education.[43] Clausewitz's strategic logic and tactical principles and Jomini's concept of lines of operation and decisive points – derived from extensive studies of Napoleon's campaigns - remain firmly rooted in the US Army and joint land warfare doctrinal and design principles.[44]

These early manifestations of operational art were inspired by the theories of annihilation (total destruction of one's enemy) and attrition (overwhelming mass at a decisive place and time). The conduct of warfare has made generational leaps in its violent expressions since the early 19th century. While the character of warfare continues to change, remarkably, the Napoleonic elements of operational art have endured.

Western operational art further evolved with Prussian Field Marshall Helmuth von Moltke (1800-1891). Historian Michael Krause credits von Moltke's genius as reconciling the debates of "short versus long war, defense versus offense,

---

42 Clausewitz, *Vom Kriege*.
    Baron Antoine Henri de Jomini, *Precis de l'Art de Guerre*, 1838. Translated in 1862 by J.B. Lippincott & Co., Philadelpha as *The Art of War*. Edited and published in 1992 with an introduction by Charles Messenger, (London: Lionel Leventhal Limited, 1992).

43 Christopher Bassford, "*Clausewitz and His Works*" courseware of the US Army War College, Carlisle Barracks in 1996 and last updated September 23, 2012. Originally published as chapter 2 of *Clausewitzean English: The Reception of Clausewitz in Britain and America*, (New York: Oxford University Press, 1994). http://www.clausewitz.com/readings/Bassford/Cworks/Works.htm (accessed on January 13, 2013).

44 The US Army "principles of war" show clear derivation from Clausewitzean principles in *Vom Kriege*. The principles are mass, objective, speed, security, surprise, maneuver, offensive, unity of command, economy of force, simplicity. Jomini's tactical precepts of lines of operation (interior and exterior), decisive points, and objective points are examples of enduring concepts still in use in US Army and Joint US doctrine and practice. These are listed in US Army Field Manual (FM) 3-0, *Operations*, February 2008, A-1 to A-4. Strategist Colin Gray analyzes the influence of Jomini and Clausewitz in modern US Army doctrine in *War, Peace and International Relations*, (New York: Routledge, 2008), 15-28.

attrition versus maneuver" within a technologically advancing society introducing advanced weaponry, expansive railroads, and improved communications methods.[45] The ascension of Prussian state and military prowess in the late 19th century is due in large part to the strategic and doctrinal vision of von Moltke. If Napoleon and von Moltke set the foundations for operational art, then the First World War (WWI) proved the catastrophic collision of European attrition warfare with modern weaponry.

WWI is less a study of operational art than it is a gruesome lesson on the limitation of tactics.[46] Tactical acumen – bound by immutable principles, hidebound rules, and vigorous execution – showed the gruesome costs required to achieve strategic objectives.[47] The Second World War (WWII) superseded this stifled operational art with a maneuver-centric German joint force decisively linking tactical success to broader campaign objectives on multiple fronts.[48] If German operational art acumen is on display in 1940-1943 by the Third Reich, so too is their strategic overreach. The case of the Third Reich reminds us that when the strategy pillar falters, so too does the skilled application of operational art. Adolph Hitler's outlandish strategic aims undermined his skilled Army and Corps Commanders in achieving feasible campaign objectives.[49]

Modern doctrinal operational design elements (Figure 3) would be quite familiar to Napoleon, Clausewitz, Jomini, von

---

45 Michael D. Krause, "Moltke and the Origins of the Operational Level of War," in *Historical Perspectives of the Operational Art*, ed. Michael D. Krause and Cody R. Phillips. (Washington DC: Center for Military History 2005), 114.
46 Brigadier General Gunter R. Roth, "Operational Thought from Schlieffen to Manstein," in *Historical Perspectives of the Operational Art*, 149-166.
47 Luttwak, Edward N., *Strategy, The Logic of War and Peace*, (Cambridge, MA: The Belknap Press of Harvard University Press), 1987, 93-112.
48 Dennis E. Showalter, "Prussian-German Operational Art, 1740-1943," in *The Evolution of Operational Art*, Krause and Phillips, 35-63.
49 Showalter, 56-58.

Moltke, and the Allied and Axis generals of WWII.[50] These thirteen doctrinal elements of operational design show the consistent threads of operational thought, language, and logic of military land power principles over the past 200 years. This consistency attests that modern operational art remains a valid guidepost for large land armies colliding on the field of battle with the singular purpose of destroying one another.

Elements of Operational Design

- Termination
- Military end state
- Objectives
- Effects
- Center of gravity
- Decisive points
- Lines of operation and lines of effort
- Direct and indirect approach
- Anticipation
- Operational reach
- Culmination
- Arranging operations
- Forces and functions

Figure 3. Elements of Operational Design, Joint Publication 5-0, III-18 Figure III-9

Not all conflict involves fielded military formations intent on the destruction of one another. A separate but related category of work studies irregular warfare.[51] In these works, the consideration of military operational art is often overshadowed by an analysis of political warfare or "armed politics." Important works in this category are *The Art of War* by Sun Tzu (500 BC), *The Seven Pillars of Wisdom* by T.E. Lawrence (1935), *On Guerilla Warfare* by Mao Tse-Tung (1937), *Modern Warfare: A French View of Counterinsurgency* by Roger Trinquier (1961), *Counterinsurgency Warfare: Theory and Practice* by David

---
50   JP 5-0, III-18.
51   JP 1-02, 159. Joint doctrine defines irregular warfare as "a violent struggle among state and non-state actors for legitimacy and influence over the relevant population(s)."

Galula (1964), *War of the Flea* by Robert Taber (1965) and *From Dictatorship to Democracy* by Gene Sharp (1993). On the whole, these works trend toward theories of exhaustion vice attrition or annihilation.[52]

USSOF strategic and tactical precepts gravitate toward these theories and practices rather than those oriented on annihilation or attrition. The majority of irregular warfare studies focus on the application of small units, teams, or individual actions that create outsized effects.[53] Often, military actions and the contemplation of military art are examined as adjuncts to political warfare. In this genre, politics, people, and psychology are the primary domains of study with military tactics as a subset of options employed to gain strategic advantages.[54]

With such divergent strategies and theories, where does one turn to examine Phase Zero operational art? Strategist and futurist Dr. John Arquilla issued an irregular warfare-centric top ten reading list to complement, if not counter-point, the United States Military Academy's top ten military classics list.[55] Arquilla calls for works examining not just the "horizontal" clashes of great powers, but also those focused on the unequal and "vertical" clashes of guerillas, bandits, and commandos. It is the latter, he claims, that "has dominated world

---

52 Theories of exhaustion center on protracted, non-decisive methods that erode the will and capacity of an adversary that seeks a decisive outcome. German historian Hans Delbruck (1848 – 1929) is credited with developing a theories of exhaustion. Examples of exhaustion theories in practice include strategies by Pericles, George Washington, Mao Tse-tung, Vo Nguyen Giap and the Afghanistan-Pakistan Taliban.
53 John Arquilla, Insurgents, *Raiders and Bandits: How Masters of Irregular Warfare Have Shaped Our World*, (Plymouth, UK: Ivan R. Dee, 2011).
54 Arquilla, *Insurgents, Raiders and Bandits*, xi-13. Arquilla examines irregular warfare in three primary categories: the small unit, guerilla warfare, and terrorism.
55 John Arquilla, "Guerilla Lit 101: Ten Books that are better than the Art of War," Foreign Policy, http://www.foreignpolicy.com/articles/2012/09/24/guerrilla_lit_101?page=0,1ptember 24, 2012 (accessed 26 September 2012). The West Point list can be accessed at the United States Military Academy Department of History website, http://www.usma.edu/history/SitePages/Home.aspx.

affairs for the past half-century – and will likely do so for the next century to come."[56] Arquilla suggests that a focused study of this irregular warfare canon is essential to understand future conflict. His argument hints at a larger question: are the time-tested, landpower-centric elements of operational art and design still valid for contemporary threats?

The Modern Need for Operational Art

Retired British General and modern war theorist Sir Rupert Smith examines this question by defining operational art in a modern warfare context. "Operational art can be understood in three ways. First, as a free, creative and original expression of the use of force and forces ... Second, in the design of the operation ... Third, in the direction of the operation to its successful conclusion. In large measure, this is an expression of the character and aptitude of the operational commander – the artist."[57] Smith's characterization of operational art contains an important qualifier critical to examining Phase Zero strategy. He cites the difficulty, if not the impossibility, of exercising operational art where no strategy exists. In a refrain familiar to Hitler's generals and to students of the Vietnam War, Smith outlines the logic and probability of failure when a strategy is muddled.

> When one party has a strategy and the other does not, the tactical acts of whatever side, being common to both, are linked inevitably to the side with the strategy – regardless of outcome. The side with the strategy is

---
56 Arquilla, *Guerilla Lit*, 1.
57 Rupert Smith, "Epilogue," in *The Evolution of Operational Art*, Olsen and van Creveld, 233.

able to exercise the operational art and, not least for want of opposition at that level, even turn its tactical failures to its advantage. It is this that gives rise to the observable phenomenon of 'winning every fight and losing the war.'[58]

In his 2005 book *The Utility of Force: The Art of War in the Modern World*, Smith asserts that industrial, state-on-state maneuver war no longer exists. Modern war is "war amongst the people."[59] Smith asserts that this paradigm shift confounds our institutions and processes that were designed and optimized for conventional, force-on-force war. Smith further claims that operational art is a casualty of this transition. The foundations upon which military strategies were built have ruptured, altering the certainty of historically sound security solutions. This dissolution of singular threats and conventional opponents further erodes our strategic logic and systems. Smith states that modern war has diluted the feasibility of clear objectives, definable theaters, cohesive command structures, and information control. In this environment, "there is little, if any evidence of the practice of the operational art, or design."[60] If Smith's troubling assertion is true, then the intellectual challenge is clear: identify and understand the modern power paradigm shifts in order to conceptualize and adopt a practice of sound, effective operational art.

58　Smith, "Epilogue," 236.
59　Rupert Smith, *The Utility of Force: The Art of War in the Modern World*, Rev. ed. (New York: Vintage Books, 2008).
60　Olsen, 241.

Designing Operational Art: Doctrinal Methods[61]

Operational art is conceptualized and created from design, or design-like thinking, that arranges military power to accomplish objectives. Design is both a process and a product using "precise and vague ideas ... systematic and chaotic thinking ... both imaginative thought and mechanical calculation."[62] Design is the conceptual thinking that precedes, then guides, detailed planning. US Army doctrine describes design as "a methodology for applying critical and creative thinking to understand, visualize, and describe complex, ill-structured problems and develop approaches to solve them."[63] Army design, now called the Army Design Methodology (ADM), is classified as an iterative approach that frames and reframes problems and environments to achieve conceptual understanding prior to detailed, programmatic planning.

Sir Rupert Smith's observations of modern strategic challenges are clearly reflected in the decade of conflict labeled the Global War on Terror (2001-2009).[64] In particular, the Iraq and Afghanistan counterinsurgencies exposed a military culture predisposed to its own organizational strengths and bound by doctrinal constructs ill-suited for irregular conflict. To overcome these propensities, the US Army led

---

61 JP 1-02. Doctrine is defined as the "fundamental principles by which military forces of elements thereof guide their actions in support of national objectives. It is authoritative but requires judgment in application," 95.
62 Bryan Lawson, *How Designers Think: The Design Process Demystified*, (Oxford: Elsevier Ltd) 2006, 4.
63 Department of the Army, Field Manual 5-0, *The Operations Process* (Washington, DC: Government Printing Office, March 2010, 3-1.
64 Al Kamen, "The End of the Global War on Terror," *The Washington Post, March 24, 2009.* http://voices.washingtonpost.com/44/2009/03/23/the_end_of_the_global_war_on_t.html, (accessed January 14, 2013). In 2009, the Obama Administration directed that the Global War on Terror moniker would be replaced with the phrase "Overseas Contingency Operations."

the joint force in developing a doctrinal process for design.⁶⁵ Recognizing that structured planning processes too often failed to stimulate creative and critical thinking, the 2010 Army planning Field Manual 5-0, *The Operations Process*, formally incorporated design into the US Army planning process.

Joint doctrine, reflecting US Army doctrinal changes, revised operational art and operational design in the 2011 version of Joint Publication 5-0, *Joint Operation Planning*. As discussed early in this chapter, joint operational art is defined as "the creative thinking used to design strategies, campaigns, and major operations and to organize and employ military force."⁶⁶ The role of operational design (Figure 4) is to "support operational art with a general methodology using elements of operational design for understanding the situation and the problem."⁶⁷ The thirteen elements of operational design are "tools which help the Joint Force Commander (JFC) and staff visualize and describe the broad operational approach."⁶⁸ In this manner, joint doctrine provides a flexible – yet definable – framework for aligning ways and means to achieve strategic ends.

---

65 Conceptually, sound design has always been a component of military planning. However, US Army doctrine considered design as an inherent task within the military planning process. In the US Army, the School for Advanced Military Studies (SAMS) at Fort Leavenworth, KS renewed the centrality of design in Army planning in the 2003-present era as a response to the modern challenges of reconciling strategy, doctrine and counterinsurgency.
66 JP 5-0, I-5.
67 JP 5-0, III-2.
68 JP 5-0, III-2.

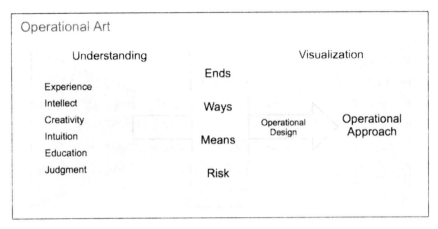

Figure 4. Operational Art, Joint Publication 5-0, III-2 Figure III-1

Operational art, once the domain of major theater commanders, is now recognized as a relevant practice at lower tactical levels. Army Doctrinal Publication 3-0, *Operations* (October 2011), defines operational art as "the pursuit of strategic objectives, in whole or in part, through the arrangement of tactical actions in time, space, and purpose."[69] In a clear break from previous doctrine, the Army declared that operational art is not "exclusive to theater and joint force commanders. Instead, it applies to any formation that must effectively arrange multiple tactical actions in time, space, and purpose to achieve strategic objectives."[70] The Director of the US Army Combined Doctrine Directorate at Fort Leavenworth, KS, stated: "One of the breakthroughs in *Army Doctrinal Publication 3-0, Operations*, (2011) is that operational art can now be applied at any level of war. Even at the battalion level, operational art is applied. This was added to Army doctrine

---

69 Department of the Army, *Army Doctrinal Publication (ADP) 3-0* (Washington DC: Government Printing Office, October 2011), 9.
70 Ibid, 9.

principally based on our OIF and OEF experiences."[71]

US Army and joint doctrine adapted to modern strategic challenges by pushing operational art down to lower levels and to smaller formations. Notably, there is significant lag time between intellectual and practical adaptations occurring in the mid-2000s and their formalization in doctrinal publications (2010-2011). However late, both US Army and Joint doctrines inculcated the themes and practices derived from the last decade of operations and produced revised definitions and methods of conceptualizing and applying operational art.

Are these doctrinal changes sufficient to contend with the modern warfare challenges articulated by Sir Rupert Smith? Are doctrinal changes that reflect Iraq and Afghanistan experiences suitable to contend with the Phase Zero environment?

The evidence suggests that despite durable doctrinal principles and recent doctrinal adjustments, the Phase Zero environment is not suited to the maneuver-centric operational art concepts and culture. The mismatch of the Phase Zero environment and modern doctrinal constructs is analogous to the struggles encountered by US land forces in Iraq and Afghanistan (2001-2006) attempting to conduct counterinsurgency warfare with maneuver-centric doctrinal education and training. Illustrative of this incongruity, Dr. Antulio J. Echevarria II of the US Army War College asserted that the 2006 US Army Field Manual 3-24 *Counterinsurgency*, "embraced a second grammar of war." Echevarria references the Clausewitzean metaphorical distinction that wars have a grammar separate from their logic. Clausewitz states:

---

[71] Author interview A18 with the US Army Director of Combined Arms Doctrine Directorate (CADD), Mr. Clint Ancker, August 29, 2012.

> War is a branch of political activity; that it is in no sense autonomous ... the only source of war is politics – the intercourse of governments and peoples; but it is apt to be assumed that war suspends that intercourse and replaces it by a wholly different condition, ruled by no law but its own. We maintain, on the contrary, that war is simply a continuation of political intercourse, with the addition of other means... Its grammar, indeed, may be its own, but not its logic. [72]

The grammar of war represents the rules that bound the realm of war and warfare: governing principles and syntax separate from its logic. The pull of grammar becomes the dominant reference for framing military operations. In effect, doctrine becomes the grammar.[73] Echevarria describes the potential insufficiencies when well-rehearsed doctrines confront ill-defined problems. "Military commanders have been substituting rules and principles literally for centuries whenever they found the twists and turns of logic too difficult to follow."[74]

In Clausewitzean terms, the thirteen elements of operational design serve as a type of grammar to create operational approaches. Notably, these thirteen elements remain intellectually associated to the land warfare principles of the Napoleonic and von Moltke eras. If understanding counterinsurgency required a second grammar of war as Echevarria states, then a proper understanding of Phase Zero may also require an amended grammar.

---

72 Carl Von Clausewitz, *Vom Kriege*, 1832. Translated, edited and published as *On War* by Michael Howard and Peter Paret (Princeton: Princeton University Press 1976), 605.

73 JP 1-02. Doctrine is "fundamental principles by which the military forces or elements thereof guide their actions in support of national objectives. It is authoritative but requires judgment in application," 90.

74 Dr. Antulio J. Echevarria II, "War's Second Grammar," Strategic Studies Institute, October 2009.

Operational Art: Modern Demands

Operational art is routinely tied to the movement of large formations and the conduct of grand scale campaigns and decisive battles. Modern security challenges that require this type of operational art are increasingly replaced with irregular threats, ill-defined conflict parameters, and multi-dimensional battle arenas. Current operational art paradigms were not created for this environment. Predictably, this makes current US planning doctrine less useful and, at times, counterproductive. New security environments demand new configurations of operational art and design.

While operational art may lag in adapting to security challenges, it is not static. The recent shifts in US Army and joint doctrine show the institutional contemplation of operational art in a changing strategic environment. If operational art is to be useful construct, it should develop a corollary for distributed, specialized operations conducted against non-state, networked threats. In addition to the threat, emerging US power projection methods, such as special operations, must be understood and accounted for in devising new operational constructs. Such manifestations of modern power – smaller in scale but not in intensity or effect – require consideration for their different expressions of operational art.

Finally, operational art becomes difficult to frame in a modern warfare environment where the principle task is not necessarily the destruction of a state's military apparatus. The modern strategic environment has ill-defined pillars with which to design, vector, and execute cohesive operational art. The challenge of devising operational art is inextricably linked to the larger challenge of identifying threats, the boundaries of a security environment, and managing the complexities of modern power paradigms.

Section II. Understanding Special Operations

In the *Aeneid*, Virgil immortalizes the idea that unorthodox thinking coupled with small, specialized teams can create startling strategic outcomes. The Trojan horse ruse by the Greeks to sack the city of Troy captures the essence of special operations: elevated risk, unconventional methods, specially selected personnel, stealth by smallness, clandestine infiltration, psychological surprise, and irregular modes of attack. As Virgil demonstrates, special operations, both in concept and execution, have deep historical traditions. This thesis examines special operations in its post-Second World War (WWII) development. In this 70-year period, special operations follows a clear trajectory as a strategic concept, capability, and force that has evolved as a distinct element of US national power.[75] This section examines the definition, development, theories, and modern applications of special operations and SOF.

Defining Special Operations

Special operations are officially defined as "operations requiring unique modes of employment, tactical techniques, equipment, and training often conducted in hostile, denied, or politically sensitive environments and characterized by the following: time sensitive, clandestine, low visibility,

---

75   The notion of special operations and special operations forces as congealing in the WWII era and beyond is best articulated by COL (Ret.) Joseph Celeski in his writings, workshops and in conversations with the author. While one can debate the origins of special operations, this study adopts Celeski's assertion.

conducted with and/or through indigenous forces, requiring regional expertise and/or a high degree of risk."[76] Special operations are conducted by special operations forces as designated by Title 10 U.S. Code, Section 167. One of nine combatant commands, USSOCOM characterizes joint SOF by the following traits: precision and scalable strike effects; ubiquitous access; regional expertise, presence and influence; C4ISR dominance; agile and unconventional logistics; force protection and survivability.[77]

USSOCOM is a unified combatant command with service-like responsibilities to resource, train, equip, and employ joint SOF. In 2005, the Secretary of Defense (SECDEF) appointed USSOCOM as the Department of Defense lead for "planning, synchronizing and, as directed, executing global operations against terrorist networks.[78] Now in its 25th year of existence, USSOCOM currently fields over 60,000 joint SOF personnel. USSOCOM comprises 4% of Department of Defense personnel and 1% of its budget.[79] By way of comparison, the United Kingdom, arguably our greatest and most capable global ally, will field a British active army totaling 82,000 soldiers by 2018.[80]

---

[76] Department of Defense, Joint Publication (JP) 1-02, *Department of Defense Dictionary of Military and Associated Terms*, (Washington DC: Government Printing Office, 08 November 2010 (as amended through 15 August 2012), 288.
[77] *Special Operations Forces Reference Manual*, Third Edition (MacDill Air Force Base, Florida: Joint Special Operations University Press) 2011, 1-4. C4ISR stands for command, control, communications, computers, intelligence, surveillance, and reconnaissance.
[78] JP 5-0, II-26.
[79] Admiral William H. McRaven, USSOCOM Posture Statement before the 112th Congress, Senate Armed Services Committee, March 6, 2012.
[80] "Strategic Defence and Security Review 2020," United Kingdom Ministry of Defence, 19 October 2010 (accessed on 26 September, 2012). http://www.army.mod.uk/news/24264.aspx

Photo 1. A US special operations soldier combat advises friendly nation forces in Africa (US Army photo, authorized use).

SOF Command Structure

The continental SOF team consists of four component commands and one sub-unified command. The commands are the United States Army Special Operations Command (USASOC), Air Force Special Operations Command (AFSOC), Naval Special Warfare Command (NAVSPECWARCOM), US Marine Corps Special Operations Command (MARSOC), and the Joint Special Operations Command (JSOC).[81]

The regional SOF joint force headquarters are called Theater Special Operations Commands (TSOC). TSOCs are sub-unified commands under USSOCOM with the

---
81 JSOU Reference manual, chapters 3-6.

majority of regionally-forward USSOF forces under the operational control (OPCON) of the Geographic Combatant Commanders. The TSOCs are Special Operations Command Europe (SOCEUR) and Special Operations Command Africa (SOCAFRICA) in Stuttgart, Germany; Special Operations Command Pacific (SOCPAC) at Camp Smith, Hawaii; Special Operations Command Central (SOCCENT) in Tampa, Florida; Special Operations Command South (SOCSOUTH) at Homestead Air Force Base; and Special Operations Command North (SOCNORTH) with the US Northern Command in Colorado Springs, Colorado.[82] Special Operations Command Korea (SOCKOR) at Camp Kim, Korea is the special operations functional component for US Forces Korea (USFK). With few exceptions, TSOCs command and control all SOF in their assigned area of responsibility.[83]

Special operations are predominately joint. The core joint task forces for special operations are the Joint Special Operations Task Force (JSOTF), Joint Military Information Support Task Force (JMISTF) and Joint Civil-Military Operations Task Force (JCMOTF). A Joint Forces Special Operations Component Command (JFSOCC) is normally established to command and control multiple joint SOF task forces.[84] Special operations core operations and activities are derived from law, joint doctrine, and USSOCOM Commander directives.[85]

---

[82] Kimberly Dozier, "US commandoes boost numbers to train Mexican forces," January 17, 2013, NBC News, accessed at http://www.msnbc.msn.com/id/50496049#.UQQlv6ViZV (accessed January 26, 2013).
[83] JSOU Reference manual, 2-14 to 2-21.
[84] JSOU Reference Manual, 2-22 to 2-29.
[85] JSOU Reference Manual, I-6 chart.

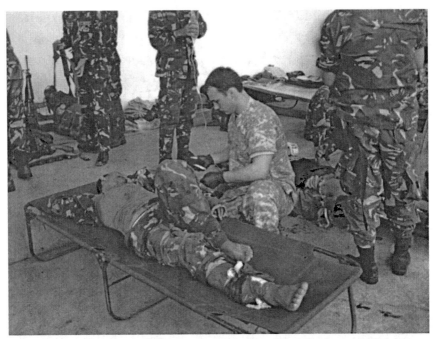

Photo 2. Special operations employ advanced medical capabilities to increase host nation medical capacity and, when prudent, deliver treatment directly to partners and populations (US Army photo, authorized use).

Historical Development of Special Operations

Modern SOF have multiple points of origin. The World War Two (WWII) era Office of Strategic Services (OSS), headed by Colonel William "Wild Bill" Donovan, is arguably the most influential strategic sponsor of today's special operations forces.[86] Donovan's OSS, designed to conduct espionage, infiltration, and subversion in the European and Pacific

---

86 Alfred H. Paddock, Jr. *US Army Special Warfare: Its Origins* (Lawrence, KS: University Press of Kansas) 2002 and George C. Chalou, *The Secrets War: The Office of Strategic Services in World War II*, (Washington DC: National Archives and Records Administration), 1992.

theaters, achieved modest success in its 1943-1945 existence. While its successes were noteworthy and promising, they were seldom strategic. OSS methods, often unpopular or misunderstood, did inspire ideas on unconventional methods of war and statecraft that would bloom in the post WWII era.[87]

Following WWII, the OSS was disbanded. This decision ultimately split the OSS roles and missions between an intelligence component, the Central Intelligence Agency,[88] and a military component, the Psychological Warfare Division.[89] It was not until 1952 that the Army resurrected an unconventional warfare capability with the creation of the 10th Special Forces Group.[90]

In the 1960s, the US Army Special Forces (USSF) became a central military component of President John F. Kennedy's anti-communist counterinsurgency strategy. Optimized for the decentralized, indigenous-focused counterinsurgency fight, USSF expanded their influence through a decade of covert, clandestine, and partnered combat operations in Vietnam. The strategic employment of USSF during the Vietnam War is still in debate. What is not in question was the utility of specially selected and trained volunteers assembled in small, elite units for the purposes of making war, managing conflict, or sustaining peace.[91] The post-Vietnam US Army force

---

87 Douglas Waller, *Wild Bill Donovan: The Spymaster who created the OSS and modern American espionage*, (New York, NY: Simon & Schuster, Inc.), 2011. Waller's recent history of William Donovan provides an outstanding perspective of the legacy of the OSS. Donovan's ideas of espionage, subversion and sabotage were dwarfed by the sheer scope and mass of WWII. His ideas grew in importance in the post WWII era where diplomacy, espionage, statecraft and small wars were more central to US strategy.
88 The National Security Act of 1947 established the Central Intelligence Agency.
89 Paddock, 38-64.
90 Paddock, 126-127.
91 Department of the Army, "Army Activities in Underdeveloped Areas Short of Declared War," Memorandum For: Secretary of the Army, 13 October 1961, authored by BG Richard G. Stilwell. Declassified on 19 October 2005. Stilwell's report, originally classified SECRET, studies the objective of the US Army in increasing its ability and capacity to increase the effectiveness of indigenous military forces in "Sublimited War."

structure recognized a prominent place for US Army Special Forces in light footprint conflicts, counterinsurgencies, and small unit military-to-military engagement.[92]

The genesis of other SOF units – US Army Rangers, Civil Affairs, Navy Special Warfare Units, Air Force Special Operations – have their own lineages outside of the OSS, also originating from the WWII era. Each of these joint SOF units is a reflection of a need within their service component for a specialized skill to enhance tactics and operations. Many niche capabilities faded or became obsolete. Others, such as the WWII-era US Navy Amphibious Scout and Raider School, survived and morphed into today's Naval Special Warfare Center. US Army Rangers followed a similar trajectory from their heroic actions at Pointe du Hoc on D-Day to become today's US Army 75th Ranger Regiment headquartered at Fort Benning, GA.

Department of Defense (DoD) counterterror and hostage rescue missions were historically conducted by ad hoc units that were built for specific missions. This method persisted until the tactical and strategic failure of OPERATION EAGLE CLAW, the failed hostage rescue attempt in Iran in 1980. Following the aborted mission, Congress ordered an investigation on the concept and execution of the mission. The commission, led by Retired Navy Admiral James Holloway III, consisted of a non-partisan board of retired general and flag officers. The Holloway Commission cited the ad hoc assemblage and poor joint interoperability of the hostage rescue team as the proximate cause of mission failure.[93] The report

---

The report singles out the Fort Bragg, NC based Special Warfare Center as the key headquarters optimized to build US Army capabilities in psychological warfare, counterinsurgency and counter-guerilla operation.

92   Paddock, 155-163.
93   Admiral James L. Holloway III, "Iran Hostage Rescue Mission Report (Unclassified)," Statement of Admiral J. L. Holloway III, USN (Ret.), Chairman, Special Operations

recommended the creation of a standing special operations joint task force singularly devoted to counter terrorism and hostage rescue. Strategically, the Holloway Commission validated the requirement for a standing joint SOF task force devoted to high-risk, direct action missions. The report resulted in creation of a standing, professionalized SOF counterterror and hostage rescue capability and laid the foundation for a consolidated SOF command. It would take seven more years after the Holloway report – and other incidents of poor joint interoperability[94] – until the legislative establishment of a strategic SOF headquarters, USSOCOM.

The 1987 Nunn-Cohen amendment to the 1986 Goldwater-Nichols act legislatively mandated a separate US special operations capability and headquarters. This formalized USSOCOM in Tampa, Florida as the consolidated special operations command. To ensure accountability and oversight, USSOCOM has specific reporting responsibilities to the US Congress.

Special Operations Theory and Doctrine

Special operations lack a single, cohesive theory that describes special operations power, strategic purpose, domain relevance, and national requirement.[95] While there is a growing body of literature and strategic contemplation of special operations, a comprehensive theory has yet to emerge.

Review Group (Washington DC: The Navy Department Library), August 23, 1980. http://www.history.navy.mil/library/online/hollowayrpt.htm, (accessed January 14, 2013).

94 United States Congress, "Public Law 99-433-Oct.1, 1986, The Goldwater-Nichols Department of Defense Reorganization Act of 1986."

95 COL (Ret.) Joseph Celeski has an unpublished manuscript that examines the current question of special operations power, strategic utility, and theory. This section reflects the author's reading of this manuscript as well as correspondence with Celeski since February, 2012.

Strategist Colin Gray observes that "although there is an abundance of literature on the unconventional derring-do of SOF, discussion of their strategic value is all but nonexistent. That is a story much in need of telling, particularly since SOF assuredly will figure with increasing prominence in the strategic history of warfare."[96]

The SOF community has recognized that the mere articulation of SOF capabilities is insufficient to inculcate special operations into US strategic thinking.[97] Lacking a recognized theoretical expression, special operations approaches, methods, and options are too often viewed as adjuncts to more traditional forms of power (air, land, maritime). Often, special operations *are* mere adjuncts to other forms of power. However, this perspective overlooks special operations as an alternative option for projecting power across multiple domains or in domain gaps, where political, social, or physical risk calculations require distinct approaches. The desire to define "special operations power" is one such outcome of the desire to reframe the language and concept of power paradigms.

In 1993, a Navy SEAL Commander named William H. McRaven developed a theory of SOF direct action operations. McRaven, now an Admiral commanding USSOCOM, argued that a small, specialized force achieving surprise could achieve "relative superiority."[98] This tactical accomplishment, combined with the proper strategic application, can achieve strategic objectives normally associated with major

---
96   Gray, Colin S. *Another Bloody Century*, (London: Phoenix, 2005), 252.
97   Author discussions with Mr. Garry Reid, Principal Deputy Assistant of Secretary of Defense for Special Operations and Low Intensity Conflict, Fort Leavenworth, KS, March 2012.
98   William H. McRaven, *Spec Ops: Case Studies in Special Operations Warfare – Theory and Practice* (New York: Ballantine Books, 1995). Of note, Clausewitz discusses the concept of "relative superiority" in Book Three, Chapter Eight of *On War*, 197.

campaigns or large military formations. Admiral McRaven's theory for commando-style raids has persisted as the fundamental theoretical touchstone for direct-action focused USSOF units. Indeed, McRaven's theory has nurtured a generation of counterterror methods, doctrine, training, and education that is executed by a globally oriented US-led counterterrorism force. Beyond McRaven's direct-action theory lies a greater theoretical challenge: developing a theory that accounts for SOF's direct *and* indirect approaches.

The Joint Special Operations University (JSOU) in Tampa, Florida, responding to the need to develop and articulate special operations theory and power, energized the discourse through USSOCOM-sponsored working groups, discussions, and writings.[99] Celeski's draft work, *An Introduction to Special Operations Power* adds theoretical components to the articulation of SOF doctrine and application.

> Special operations are those activities of unorthodox nature applied to increase the 'fog and friction of war' on our adversaries (a form of political warfare – Special Operations Forces medium); if applied correctly, they can achieve strategic effects in the arena of grand strategy, independent of operational art associated with campaign plans and objectives. In military campaigns, these actions are applied in cooperation with conventional force maneuver (or its inverse, conventional forces supporting SOF maneuver) to achieve

---

99 As part of this discourse, the author reviewed two as-yet unpublished manuscripts written by COL (Ret.) Celeski in 2011-12. The first is titled *Political Warfare*. The second, referenced in chapter one, is *An Introduction to Special Operations Power: Origins, Concepts, and Application*. COL (Ret.) Celeski used the ideas in these manuscripts as the lead-in to professional discussions and professional development programs conducted at the Command and General Staff College (CGSC), Fort Leavenworth, KS in February - March 2011.

military objectives. Outside of war, special operations activities are becoming more important in shaping environments for deterrence of potential adversaries and cooperating with our allies as a primary diplomatic and foreign policy tool of the United States.[100]

In Celeski's analyses, special operations is theoretically analyzed across all domains (horizontally) and across the strategic-operational-tactical spectrum (vertically). This approach reveals two key challenges for the formulation of a special operations theory. The first challenge is identifying special operations in relation to a domain. The second challenge is reconciling – within one theory - the two opposing natures of special operations: indirect and direct.

Domains are the cognitive categories that frame expressions of military power. The Department of Defense recognizes five domains: air, land, maritime, space, and cyber.[101] Accordingly, forces and capabilities are defined by their utility in influencing, controlling, or dominating a recognized domain. The domain paradigm provides clear expressions for land forces (land), air forces (air), and naval forces (maritime). Where do special operations fall into the domain paradigm?

Special operations are considered cross-domain, multi-domain, or best suited to operate in domain gaps. Unlike airpower, land power, and sea power, special operations are generally not designed to dominate a domain (or to do so temporally). Rather, SOF seek to create advantages within domains by generating or reducing frictions. SOF can exploit or counter adversary actions in domain gaps.[102] Special

---

100 Joseph Celeski, *Special Operations Theory and Power*, unpublished monograph, used by permission of the author, 2013.
101 JP 5-0, xx.
102 Celeski, conversation with author in March 2012.

operations are well-suited to apply indirect forms of power: humanitarian assistance, moral diplomacy, remote-population engagements, and influence activities. However untidy this broad description may be regarding the application of special operations power, it reflects the complex strategic environment in which SOF are well suited to operate: nebulous domain seams and sensitive locales.

Recent scholarship from the United States Army Special Operations Command (USASOC) promotes the idea that SOF operate in an as-yet unrecognized domain: the human domain.[103] The human domain represents an emerging concept that defies simple categorization but merits consideration as the US adjusts the paradigm of modern military power projection.[104]

Instead of the domain quandary inhibiting the understanding of special operations, the opposite logic must be considered: domains may be inadequate paradigms to frame our conceptualization and projection of power. If so, the strategic utility of SOF is not a puzzle to be put together. Instead, we must reconsider our view of domains as a useful expression of national power. The Arab Spring and social media revolutions in 2011 are recent examples where existing domain paradigms fell short in categorizing citizen-centric, social media-enabled power schemes that toppled longstanding regimes in the Middle East and North Africa.

The second challenge in formulating a special operations

---

103 USASOC introduced the term 'human domain' in the 2012 draft of Army Doctrinal Publication 3-05, *Army Special Operations*. On 15 May 2012, the Doctrine 2015 General Officer Review Board delayed the inclusion of the term in Doctrine 2015 publications to allow for the further development and consideration of the idea. Sacolick and Grigsby, "Special Operations/Conventional Force Interdependence," 39-42.

104 Department of the Army, TRADOC PAM 525-3-0, *The Army Capstone Concept*, 19 December 2012, (Washington DC: Government Printing Office, 2012). The concept states "current doctrine does not adequately address the moral, social, cognitive, and physical aspects of human populations in conflict," 15.

theory is the dual nature of special operations. Often called "direct and indirect," these capabilities range from the shock power of specialized counterterrorism raids to hidden-hand psychological influence operations oriented toward cognitive targets. With such a disparate range of methods, combining special operations into a single paradigm becomes problematic.

To improve the strategic understanding of special operations, USASOC introduced revised doctrinal language with a clear declaration of special operations dualism. Issued in August 2012, Army Doctrinal Publication 3-05, *Special Operations*, describes special operations in two distinct categories: special warfare and surgical strike.

> *Special warfare* is the execution of activities that involve a combination of lethal and nonlethal actions taken by a specially trained and educated force that has a deep understanding of cultures and foreign language, proficiency in small-unit tactics, and the ability to build and fight alongside indigenous combat formations in a permissive, uncertain, or hostile environment. Special warfare is an umbrella term that represents special operations forces conducting combinations of unconventional warfare, foreign internal defense and/or counterinsurgency through and with indigenous forces or personnel in politically sensitive and/or hostile environments.[105]

Special warfare represents the classically "by, through, and with" special operations missions most often associated

---

[105] Department of the Army, *Army Doctrinal Publication (ADP) 3-05, Special Operations*, (Washington DC: Government Printing Office, October 2011), 9.

with US Army Special Forces. The psychological and civil affairs components – the nonlethal actions – that typify SOF indirect approaches are also categorized as applications of special warfare. Surgical strike represents the special operations capability most known as direct-action raids, hostage rescue, or other deep penetration, high-risk operations.

> *Surgical strike* is the execution of activities in a precise manner that employ special operations forces in hostile, denied, or politically sensitive environments to seize, destroy, capture, exploit, recover or damage designated targets, or influence threats. Executed unilaterally or collaboratively, surgical strike extends operational reach and influence by engaging global targets discriminately and precisely. Surgical strike is not always intended to be an isolated activity; it is executed to shape the operational environment or influence select target audiences in support of larger strategic interests.[106]

Special warfare and surgical strike represent SOF expressions and options for the broader, joint, operational approaches described as direct and indirect approaches.[107]

The codification of SOF dualism, new to ARSOF doctrine, is currently under consideration for joint SOF doctrine.[108] This paradigm attempts a practical and useful construct to explain the purpose and essence of special operations, without losing

---

106 ADP 3-05, *Special Operations*, 10.
107 JP 5-0, "The approach is the manner in which a commander contends with a center of gravity (CoG). A direct approach attacks an enemy's CoG or principal strength by applying power directly against it … An indirect approach attacks the enemy's CoG by applying combat power against a series of decisive points that lead to the defeat of the CoG while avoiding enemy strength," III-31 to III-32.
108 Author visit to USSOCOM, December 2012.

the nuances inherent in each approach.

Army Doctrinal Publication 3-05, *Special Operations*, also explains special operations in relation to the US Army land power doctrinal construct of Army Doctrinal Publication 3-0, *Unified Land Operations*. Though no single theory of special operations is widely accepted, the strategic utility of special operations, in both direct and indirect approaches, has an appreciable body of history, theory, doctrine, and practice to consider revised operational art constructs.

Photo 3. Special operations Cultural Support Teams (CSTs) are small, female-led teams with specialized skills to engage local populations (US Army photo, authorized use).

Special Operations: Contemporary Application

Examining special operations in Phase Zero should categorically exclude the past ten years of SOF Iraq and Afghanistan experiences that occurred in later phases: Phase Four (Stabilization) and Phase Five (Enable Civil Authorities).[109] However, to understand the contemporary application of special operations in any phase, the transformative effects of the Afghanistan and Iraq experiences must be acknowledged. In the past decade, five critical transformations shaped modern special operations and are central to understanding the USSOF contribution to Phase Zero environments.

The first transformation regards one of USSOF's core missions: unconventional warfare (UW). Both the Afghanistan and Iraq wars were initiated with classic unconventional warfare campaigns combined with modern air power. Unconventional warfare is defined "as activities conducted to enable a resistance movement, or insurgency, to coerce, disrupt or overthrow a government or occupying power by operating through or with an underground, auxiliary or guerilla force in a denied area."[110] The October 2001 SOF and Central Intelligence Agency campaign with the Northern Alliance, combined with joint air power (and supported and enabled by other services), reintroduced unconventional warfare as an effective strategic alternative to major land operations.[111] In 2003, when Turkey denied US access to the 4th Infantry Division for the invasion of Northern Iraq, a SOF unconventional warfare task force executed a doctrinal unconventional

---

109 The joint doctrine Phasing model in JP 5-0, Joint Operation Planning, are shape (phase 0), deter (phase I), seize initiative (phase II), dominate (phase III), stabilize (phase IV), and enable civil authority (phase V), III-44.
110 JP 3-05, II-9.
111 Charles H. Briscoe and Richard L. Kiper, *Weapon of Choice: U.S. Army Special Operations Forces in Afghanistan*, (Fort Leavenworth, KS: Combat Studies Institute, 2003).

warfare campaign with Kurdish irregulars penetrating the defensive line of multiple conventional Iraqi divisions and opening up the northern approaches into Iraq.[112] The use of unconventional warfare in maneuver campaigns with irregulars was again demonstrated as a differentiated, if not a preferable, strategic option. Unconventional warfare – coupled with precision air power - gained a new appreciation from policy makers, strategists, and military leaders as the correct application of small, specialized forces in achieving major campaign objectives.

The second transformation was the growth of US counterterrorism (CT) capabilities. Operation Enduring Freedom-Afghanistan (OEF-A) and Operation Iraqi Freedom (OIF) transformed USSOF and interagency counterterrorism infrastructure from a niche capability into a global network of intelligence, communications, targeting, and unprecedented operational reach. Counterterror forces, formerly cloistered on small bases awaiting presidential authorizations, are now forward deployed in multiple COCOMs, executing steady-state operations. This has resulted in the refinement of targeting methodologies such as "Find, Fix, Finish, Exploit, Analyze, Disseminate" or F3EAD.[113] Counterterror forces have now become inseparable from interagency partners from intelligence, law enforcement, border, customs, and other federal agencies with counterterror responsibilities. Organizationally, the progression of counterterror operations has transformed USSOF

---

112 Charles H. Briscoe, Kennethy Finlayson, Robert W. Jones Jr., Cherilyn A. Walley, A. Dwayne Aaron, Michael R. Mullins, and James Schroder, *All Roads Lead to Baghdad: Army Special Forces in Iraq*, (Boulder, CO: Paladin Press, 2007).
113 Charles Faint and Michael Harris, "F3EAD: Ops/Intel Fusion 'Feeds' the SOF Targeting Process," Small Wars Journal, January 31, 2012, http://smallwarsjournal.com/jrnl/art/f3ead-opsintel-fusion-"feeds"-the-sof-targeting-process (accessed October 4, 2012). Faint and Harris explain the origins and utility of the F3EAD methodology. This article is an excellent analysis of doctrinal targeting constructs.

units. As a result, USSOF developed advanced capabilities and methods for operations against terror threats and networks. Finally, the secretive counterterrorism cloak has been lifted, to practical levels, to capitalize on shared understanding and collaboration from all partners, agencies, and forces, in attacking terror networks.[114]

Photo 4. Special operations combat advisors conferring with partnered force soldiers (US Army Photo, authorized use).

Combat Foreign Internal Defense (FID) is the third critical transformation.[115] Ten years of sustained partnership with Iraq and Afghan forces, often in pitched battle, developed

---

114 Stanley A. McChrystal, "Becoming the Enemy: To win in Afghanistan, we need to fight more like the enemy," *Foreign Policy*, March/April 2011, 66-70.
115 JP 1-02. Foreign Internal Defense is defined as "participation of civilian and military agencies of a government in any of the action programs taken by another government or other designated organization to free and protect its society from subversion, lawlessness, insurgency, terrorism and other threats to security," 121. "Combat FID" is an informal term used the by US Army Special Forces Regiment used to describe environments such as Afghanistan where FID is conducted in environments characterized by sustained combat operations.

new modes of combat FID (partnered operations). Bottom-up intelligence gained from Army, Navy, and Marine Special Operations units became a prized commodity in counterinsurgency (COIN) environments. SOF developments in navigating, collecting, and operationalizing intelligence in the human domain permeated the force structure, training pipeline, and educational requirements. Combat FID advances occurred as SOF partnered with all manner of foreign security forces: special operations, regular army, border control, infrastructure security, police, intelligence, irregular forces, and special activities.[116] Today's SOF methodologies for partnering, training, targeting, and operating with foreign security forces are indelibly stamped with the introspective lessons of the Afghanistan and Iraq campaigns.

The fourth transformation is the development of Village Stability Operations and Afghan Local Police (VSO-ALP). In modes of warfare without uniformed enemies to decisively defeat, the population becomes the object to win. The development of VSO in Afghanistan resurrected a Vietnam-era tactic of developing community defense mechanisms.[117] VSO-ALP answered a strategic need: the requirement to generate bottom-up momentum to connect to and compliment faltering top-down programs. VSO-ALP capitalized on the SOF core competencies of small footprints, unobtrusive force protection, skilled engagements with locals, the use of small-scale development, and raising security forces from unskilled irregulars. In 2012, USSOF sponsored over 60 VSO-ALP sites across Afghanistan.[118] The strategic effect

---

116   Dave Butler, "Lights Out: ARSOF Reflect on 8 Years in Iraq," *Special Warfare*, Volume 25, issue 1, January-March 2012 (Fort Bragg, NC), 28-34.
117   Seth Jones, "Going Local: The Key to Afghanistan," *The Wall Street Journal*, August 7, 2009.
118   Lisa Saum-Manning, "VSO/ALP: Comparing Past and Current Challenges to Afghan Local Defense," December 27, 2012, Small Wars Journal, http://smallwarsjournal.

of VSO-ALP has yet to be judged. While the external effect is still under debate, the emergence of VSO-ALP stimulated change internally in USSOF.[119] The successes and failures of this difficult mission generated significant intellectual introspection about the optimum employment of SOF within contested populations.[120]

Photo 5. A special operations soldier patrols with an Afghan local policement (US Army photo, authorized use).

The fifth and final transformation is organizational – the development of the Special Operations Joint Task Force (SOJTF).[121] The maturation and growth of SOF led to USSOCOM's introduction of the SOJTF concept in 2011 joint

---

com/jrnl/art/vsoalp, accessed on January 3, 2012.
119 Author experience with the early establishment of Village Stability Operations – Afghan Local Police programs from 2009 to 2013.
120 Author observation spanning from 2009 to 2013.
121 JP 3-05, Joint Special Operations, III-10.

doctrine.[122] The SOJTF construct organizes special operations as the Combined Joint Task Force (CJTF) when special operations approaches are the lead lines of effort in major campaigns or theaters.[123] The current SOJTF in Afghanistan has combined the disciplines of counterterrorism, foreign internal defense, VSO-ALP, Military Information Support Operations (MISO), and multinational special operations into one fused effort, commanded by a single headquarters. This evolution of the SOJTF blends the dual capabilities of SOF with the capacity to command non-SOF formations. The SOJTF expands the Joint Task Force options, particularly when major operations are dominated by SOF core competencies. This evolution has clear advantages for future Phase Zero environments where the blending of counterterrorism, partner capacity building, and psychological and civil actions can potentially be directed by a single headquarters.

The above five USSOF transformations typify the modern modalities, capabilities, and mindsets in today's USSOF formations. These changes and experiences are certain to shape the thinking, organization, and efficacy of USSOF involvement in Phase Zero environments.

Summary

Special operations, in spirit and principle, are relatively unchanged from the trojan horse ruse employed by the Greeks 2,500 ago. However, the post WWII-era of modern special operations illustrates a force adapting to address capability shortfalls and domain gaps with highly specialized, low-density skills. Since WWII, special operations have developed

---

122 JP 3-05.
123 JP 3-05, III-10 to III-12.

an organizational essence defined by its theory, doctrine, capabilities, culture, and the emerging strategic recognition of "special operations power." The last twelve years have witnessed tremendous growth in special operations missions, force structure, tactical prowess, and operational utility.[124] This trajectory of modern special operations provides context useful to conceptualizing new applications of special operations in Phase Zero environments.

Section III. Understanding Phase Zero

This section provides definitions, histories, theories, and modern environment of Phase Zero. Combining all steady-state military actions under the rubric of "Phase Zero" risks framing all US foreign security affairs solely through a martial lens. This section attempts to avoid this trap by equally examining the diplomatic and military aspects of pre-crisis environments.

Definition of Phase Zero

Phase Zero is the slang descriptor for both the actions taken and the environment involved in maintaining US access and influence through foreign engagements with means and methods below the threshold of war.[125] Phase Zero is the first phase of the joint doctrine phasing model for operations.[126]

---

124 Since September 11, 2001, USSOCOM has doubled in size from 30,000 service personnel to nearly 67,000 today. In the same period, the budget more than quadrupled from $2.2 billion in 2001 to over $10.5 billion in 2012.
125 Author definition. Phase 0 (Shape) has a joint doctrinal definition derived from the joint phasing model. The shift from "Phase 0" to "Phase Zero" in this study connotes a stand-alone description of activities that occur in a peacetime environment and not solely as a "preparation" phase for joint operations. Joint doctrine is flexible on the phasing and is not prescriptive about following chronological phases.
126 JP 5-0, III-42. The definitions is as follows: "Shape (Phase 0). Joint and multinational

The six joint phases are shaping (phase 0), deter (phase I), seize the initiative (phase II), dominate (phase III), stabilize (phase IV), and enable civil authority (phase V).[127] To preserve flexibility in operational design, joint doctrine states, "the six-phase model is not intended to be a universally prescriptive template for all conceivable joint operations and may be tailored to the character and duration of the operation to which it applies" (Figure 5).[128] The doctrine specifies that "operations and activities in the *shape* phase normally are outlined in Theater Campaign Plans (TCPs) and those in the remaining phases are outlined in Joint Strategic-Capabilities Plan (JSCP) directed contingency plans."[129] Joint doctrine makes a critical distinction by separating *shape* (in Phase Zero) as principally directed by Theater Campaign Plans while the remaining five phases are directed by contingency (wartime) plans.[130] This distinction gives Phase Zero clear properties that are related to, but separate from, pure military contingency or wartime operations.

---

operations – inclusive of normal and routine military activities- and various interagency activities are performed to dissuade or deter potential adversaries and to assume or solidify relationships with friends and allies."
127 JP 5-0, III-38.
128 JP 5-0, III-41.
129 JP 5-0, III-42.
130 The word "principally" is used here because joint doctrine states that "While most shaping activities are contained in the TCP, contingency plans may include shaping activities that must be accomplished to support an operation." JP 5-0, III-42. In this case, Theater Campaign Plans (TCPs) describe Phase Zero operations, actions and activities and contingency plans represent staffed, approved and reasonably rehearsed wartime military contingency operations.

BACKGROUND

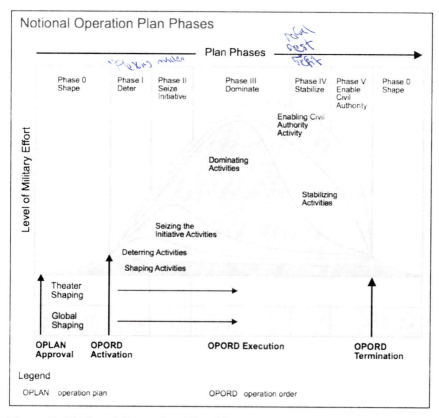

Figure 5. Notional Operation Plan Phases, Joint Publication 5-0, III-39 Figure III-16

Shape is doctrinally defined as "joint and multinational operations – inclusive of normal and routine military activities – and various interagency activities performed to dissuade or deter potential adversaries and to assure or solidify relationships with friends and allies."[131] Joint doctrine nomenclature combines the verb shape with its numerical phase, Phase 0. Though incorporated into one term, *shape* is clearly the principle activity with *phase 0* in parenthesis to denote the

---

131 JP 5-0, xxiii.

chronological phasing scheme where shaping actions occur. To provide clarity in this study, the term Phase Zero is used in place of *shape* to prevent confusion about shaping operations, which take many forms and occur in all phases. Throughout this study, *Phase Zero* is the rubric used to describe both the actions (engagement, diplomacy) and the environment (pre-crisis, peacetime). Phase Zero operations, actions, and activities describe military engagements not in a wartime environment or "left of the Joint Task Force."[132]

In the peacetime environment of Phase Zero, diplomacy is the lead discipline for the attainment of US foreign policy interests. The lead agency for the US is the Department of State (DoS). Phase Zero military affairs, however robust, should occur within the diplomatic frameworks and operating parameters set by Department of State. Diplomacy is "the conduct of international relations by negotiation rather than by force, propaganda, or recourse to law, and by other peaceful means (such as gathering information or promoting goodwill) which are either directly or indirectly designed to promote negotiation."[133] Diplomacy aims to achieve national policy by employing dialogue, negotiation, custom, law, and other methods of statecraft: a perpetual continuum of lawful statesmanship and civil discourse. Culturally and programmatically, the US Department of State does not use a strict phasing construct for their activities. In effect, all diplomatic relations with nations occur in a Phase Zero environment until such time as the hostilities escalate requiring more intrusive actions (Phase I, deter) or deliberate combat (Phase II/III, seize the initiative/dominate).

---

132 US Army Special Operations Command (USASOC), unclassified command brief (power point), June 2012.
133 G.R. Berridge, *Diplomacy: Theory and Practice*, (Hertfordshire, UK: Prentice Hall), 1995, 1.

Historical Development of Phase Zero

Following WWII, the US adopted the Unified Command Plan (UCP) structure to designate regional responsibilities and authorities for US global engagement.[134] The Unified Command Plan method enabled geographic combatant commanders (GCCs) to conduct steady-state military-to-military engagements in pursuit of sustained relational contact with adversaries, friends, and allies.

Under the Unified Command Plan construct, military shaping activities fell into two categories: preparation activities in support of major (wartime) contingency plans and military engagements conducted for the purposes of sustaining influence, relations, and access to countries deemed critical to national interests. The strategic focus of the military remained the first category: a force designed for decisive combat operations against foreign military forces.[135]

In 1961, President John F. Kennedy initiated a shift in strategy to small wars and sustained influence engagements. This was a marked shift from the conventional focus of US military leadership.[136] This shift was codified in the 1961 Foreign Assistance Act (FAA) that legislated a new foreign assistance program structure and process.[137] The 1961 FAA created the

---

134 Ronald H. Cole, Walter S. Poole, James F. Schnabel, Robert J. Watson, Willard J. Webb, *The History of the Unified Command Plan, 1946-1993*, (Washington DC, Joint History Office, Office of the Chairman of the Joint Chiefs of Staff, February 1995).

135 Dale R. Herspring, *The Pentagon and the Presidency: Civil-Military Relations from FDR to George W. Bush*, (Lawrence, Kansas: The University Press of Kansas, 2005), 52-118. Herspring's examination of civil-military relations in the Truman and Eisenhower eras provides the political backdrop to the formation and debate of the post WWII US armed forces purpose and composition. The US force structure was oriented on large scale, decisive conventional operations against a peer or near-peer competitor.

136 Herspring, 140-143.

137 Committee on International Relations and Committee on Foreign Relations, *Legislation on Foreign Relations Through 2002*, (Washington DC: US Government Printing Office, 2003). Current legislation and related executive orders. Accessed on http://transition.usaid.gov/policy/ads/faa.pdf on November 12, 2012. The origins of USAID date back

dual engagement strategy of military and non-military aid, enabling US limited war methods to stop Soviet communist expansion. The FAA created the US Agency for International Development (USAID) to carry out the development and aid missions. Military aid to foreign nations (labeled "security assistance") was controlled and funded by the Department of State. The Department of State was responsible for administering the security assistance programs.[138] Concomitant to the 1961 Foreign Assistance Act, the Kennedy Administration championed and expanded US Army Special Forces as the military force ideally suited to combat communist insurgencies.[139] US Special Operations Forces, as a combat force and with a distinct operational approach, reached a high water mark of influence under the Kennedy administration. Following the Vietnam War, successive presidential administrations and national security emphases became focused on major theater nuclear war and conventional conflict, principally in Europe.[140] This security environment characterizes the Cold War period from the US withdrawal from Vietnam War (1973) up to the dissolution of the Soviet Union (1991).[141]

The post Cold War world demanded new security concepts, approaches, and posture. The 1995 National Security Strategy subtitled, "A Strategy of Engagement and

---

to the post-WWII era. Building on the success of the Marshall Plan, President Harry S. Truman proposed an international development assistance program in his 1949 inaugural address. Several precursor organizations were created between 1950 and the 1961 to support US aid abroad.

138 Foreign Relations Committee, Legislation on Foreign Relations Through 2002.
139 Video: "JFK on Special Forces" June 6, 1962 address to the United States Military Academy graduating class, http://www.youtube.com/watch?v=7FVrpiG7haE. Accessed on November 12, 2012.
Department of the Army, "Army Activities in Underdeveloped Areas Short of Declared War," 1961. Paddock, 33.
140 Andrew J. Birtle, *U.S. Army Counterinsurgency and Contingency Operations Doctrine 1942-1976,"* (Washington DC: Center of Military History, 2006), 477-495.
141 Allan R. Millett and Peter Maslowski, *For the Common Defense: A Military History of the United States of America*, (New York: The Free Press, 1984), 542-587.

Enlargement" codified the use of peacetime engagement as one of three lines of effort (peacetime engagement, deterrence and conflict prevention, fight, and win) to achieve US strategic objectives.[142] During the next ten years (1995-2005), the naming conventions for theater engagements shifted in name but not intent. The names include *Theater Engagement Planning* (1998), *Theater Cooperation* (2005), and the current moniker, *Security Cooperation*.[143]

Following the September 11, 2001 terror attacks on the United States, Secretary of Defense Donald Rumsfeld added a new dimension to engagement:

> Secretary Rumsfeld has directed his combatant commanders and others in the Department to move beyond the relatively unfocused practice of "engagement" – which sometimes amounted to little more than "showing the flag" abroad – and toward a more specific and practical set of goals to lay the foundation for our partners working with us in defense endeavors in the future.[144]

For the armed forces, this directive forced a recalculation of all aspects of engagement: countries, regions, partners, duration, frequency, authorities, and outcomes. Rumsfeld's directive generated the impetus to operationalize engagement with a clear focus on terror threats. His guidance expanded

---

142 The White House, "A National Security Strategy for Engagement and Enlargement," 1995.
143 Department of Defense Directive (DoDD) 5132.03, 24 OCT 2008. Prior names are found in the Chairman of the Joint Chiefs of Staff (CJCS) 3113.01A, 1998.
144 Douglas J. Feith, Transformation and Security Cooperation, remarks by Under Secretary of Defense for Policy Douglas J. Feith, Washington DC, September 8, 2004 available at http://defenselink.mil/Speeches/Speech.aspx?SpeechID=145, accessed on November 12, 2012.

engagement beyond its original dual purpose of setting the theater for the introduction of large, conventional Army forces and to project US presence abroad. Consistent with the offensive-spirited Global War on Terror (GWOT), Rumsfeld's directive began the shift to more muscular and synchronized engagements with improved strategic links among intelligence, threats, foreign capabilities, and other overseas activities.[145]

Rumsfeld's concept, however visionary, met with a web of legal, authoritative, funding, and resourcing challenges.[146] Achieving Rumsfeld's idealized Phase Zero engagements required significant and controversial legislative and structural revisions. Few revisions were implemented.[147] In late 2004, the US national security challenge of integrating the interagency in Phase Zero countries was an important, but clearly secondary effort. During this period the US was grappling with bigger problems in a different phase: Phase Four (stability) in Iraq and, to a lesser extent, Afghanistan.

The increasingly costly and ineffective US war effort in Iraq in 2004-2005 revealed that the US military and civil service were poorly prepared to conduct large-scale stability operations.[148] To address this shortcoming, the Bush Administration issued National Security Presidential Directive (NSPD) 44, *Management of Interagency Efforts Concerning Reconstruction and Stabilization* (December 7, 2005). In the same week

---

145 "Embassies as Command Posts in the Anti-Terror Campaign: A Report to the Committee on Foreign Relations," United States Senate, Richard Lugar, Chairman, 109th Congress, (Washington DC: US Government Printing Office), December 15, 2006, 18.
146 Project for National Security Reform, "Ensuring Security in an Unpredictable World: The Urgent Need for National Security Reform," report, July 2008, www.pnsr.org (accessed January 15, 2013).
147 Embassies as Command Posts in the Anti-Terror Campaign, 1-18.
148 Many official government and private media accounts details the US' insufficiencies in contending with a growing counterinsurgency in Iraq. For an excellent journalistic perspective, see Tom Ricks' *Fiasco: The American Military Adventure in Iraq*, (New York, New York: Penguin Books, 2006).

(November 28, 2005), the Department of Defense issued Department of Defense Directive (DoDD) 3000.05, *Military Support for Stability, Security Transition, and Reconstruction (SSTR) Operations*, directing that stability operations are "a core US military mission."[149] These directives provided a clear mandate: the major intellectual, legislative, and resource energies of the US government were to be focused on stability operations. In this environment, US government policies and resources focused on *post*-war issues under the catch-all title of counterinsurgency (COIN). This focus left a smaller national effort dedicated to *pre*-war or *non*-war shaping activities and operations in Phase Zero. While a cottage industry grew around counterinsurgency, the troubleshooting of Phase Zero problems was not undertaken in Washington DC, but at the individual embassy level.[150]

The Development of Diplomacy

The role of diplomacy is central to understanding how the US conducts peacetime engagement in foreign nations. A brief history and background of diplomacy is provided in order to provide a view of engagement that involves more than the military instrument of power.

Diplomacy is a "continuous rather than an episodic process"[151] that formalizes relations and communications between nations, states, and peoples. Former US Secretary of State and Nobel laureate Henry Kissinger called diplomacy the "art of restraining power."[152] Political scientist G.R.

---

[149] Department of Defense Directive (DoDD) 3000.5, 28 NOV 2005, USD (P), Military Support for Stability, Security, Transition, and Reconstruction (SSTR) Operations.
[150] Embassies as Command Posts in the Anti-Terror Campaign, 1-18.
[151] G.R. Berridge, *Diplomacy: Theory and Practice*, (Hertfordshire, UK: Prentice Hall), 1995, 1.
[152] This quote is widely attributed to Kissinger thought its exact origin is unclear.

Berridge stated, "The chief purpose of diplomacy has always been to enable states to secure the objectives of their foreign policies without resorting to force, propaganda, or law – in short, by lobbying and negotiation"[153]

In the fifteenth and sixteenth centuries, state relations contained an essence of cunning and deceit loosely bound by protocols. This pre-Westphalian diplomacy is typified in Niccolo Machiavelli's *The Prince* (1469-1527) and the shrewd statecraft of France's proto-Prime Minister Cardinal Richelieu (1585-1642). In the seventeenth and eighteenth centuries, the French transformed this Italian system into a more professionalized arrangement.[154] Berridge cites the French contributions to diplomacy in the areas of professional qualifications, resident diplomats, ceremonial and protocol norms, secrecy in negotiations, and spreading French as the required language of diplomatic discourse. In this system, most diplomatic efforts remained bilateral.

In the twentieth century, multi-lateral diplomacy grew in importance and sophistication. In the spirit of the Congress of Vienna (1815), the benefits of international, multilateral conventions came to be recognized as a potential hedge against future Napoleonic-type imperialism.[155] The high aspirations of multilateral diplomacy in the twentieth century are exemplified by the League of Nations (1919-1946) and its reformed successor, the United Nations. The European Union, crafted in spirit to prevent the hegemonic rise of a reborn Germany, similarly provides a binding political framework to normalize

---

[153] Geoff Berridge, "Geoff Berridge on why we need diplomats," blog interview by Georgie Day, The Browser: Writing worth reading, posted on http://thebrowser.com/interviews/geoff-berridge-on-why-we-need-diplomats (accessed December 12, 2012).
[154] Berridge, 2-9.
[155] Colin S. Gray, *War, Peace and International Relations: An Introduction to Strategic History*, (New York: Routledge, 2008), 31-61.

and advance interstate relations.[156]

Today's US Department of State has over 256 foreign postings at embassies, consulates, and other diplomatic missions in over 180 countries.[157] The US diplomatic corps, both state employees and contractors, is estimated to be about 24,000, excluding foreign nationals.[158]

Despite its size and global infrastructure, US diplomacy is best expressed by its forms: mediation, negotiation, policymaking, policy-interpreting, social relations, oversight, and protection of a nation's interests overseas, promotion of US ideals, and, importantly, direct representation of the President of the United States.[159] US embassies manage the full complement of US interests across the diplomatic, information, military, and economic (DIME) spectrum employing attaches, political officers, economic officers, public diplomacy staffs, and other specialists required to advance US interests.[160]

Theories of Diplomacy

Like special operations, diplomacy is not guided by a single, unified theory. The theoretical basis of diplomacy remains the belief that cooperative discourse and civil mechanisms are preferable methods of pursuing interests than violent means. In the pursuit of strategy, the art of diplomacy is a practical, continuous affair of negotiation and compromise. In practice,

---

156 Author visit to the US Delegation to the European Union, Brussels, Belgium, October 2012.
157 US Department of State, official website, www.us.state.gov.
158 Estimates vary on the actual number of DoS government employees and contractors. There is no official number that is published for public record. This estimate was taken from open source information gained from a field visit to the Department of State in July 2012.
159 This list of activities was compiled from the information available on the US Department of State website, www.state.gov, accessed November 12, 2012.
160 US Department of State, official website, www.state.gov.

this approach is captured by Henry Mintzberg's concept of *emergent* strategies.

Emergent strategies result in outcomes – not always intended – when deliberate strategies are churned within the dynamic interplay of policy, relationships, country interests, economics, security, and human relations.[161] The give-and-take nature of diplomatic relations is likely to produce emergent strategies continuously seeking advantages rather than definitive end-states.[162] Diplomatically, strategic success is often viewed as continuation rather than culmination.[163] Given this theoretical and practical nature of diplomacy, one can expect inevitable tensions with the decisive action favored by many military strategists and practitioners.

Militaries implement *deliberate* plans and programs in phasing constructs designed to maximize US security advantages short of, and including, war. In Phase Zero, militaries also seek a position of continuing advantage, both to sustain the peace and to prepare for war. In pursuit of greater advantages, militaries view risk-reward calculations through a unique lens. Military plans frame the world through threats and vulnerabilities that are managed and mitigated by detailed plans and preparations. The military art is to reduce risks, shrink vulnerabilities, and implement preparations prudent to contend with probable threats. Risks are accepted, even expected, if the rewards are clear and attainable.[164]

---

161 Henry Mintzberg, *The Rise and Fall of Strategic Planning: Reconceiving Roles for Planning, Plans, Planners*, (New York, New York: The Free Press, 1994), 24. Mintzberg's description of deliberate and emergent strategies has corollaries to the logistically-detailed military strategies and plans (deliberate) and diplomatic strategies based on objectives and aspirations viewed through patterns of diplomatic relationships and outcomes. In both cases, the result is emergent strategies where unintended consequences shape the actual strategic outcomes.

162 Everett Carl Dolman, *Pure Strategy: Power and Principle in the Space and Information Age*, (New York, New York: Frank Cass Publishers, 2005), 6.

163 Dolman, 4.

164 JP 5-0, *Joint Operations Planning*, contains the doctrinal templates, logic, plans scheme,

Conversely, a diplomat may not share the risk-reward viewpoint of the military strategist singularly focused on security threats. The over-militarization of state affairs invites risks that can reframe state relations.[165] However prudent the threat mitigation plans may be, military options can color the actions of the statesman in a martial light, or worse, rebuke the supremacy of civil discourse. The tensions of emergent and deliberate strategies are exacerbated by differing views of risk-reward calculations. A USSOF officer with experiences working in embassies framed these perspectives: "A day where the status quo of peace is maintained is a good day for the State Department whereas a day where nothing changes is viewed as stagnant by the Defense Department."[166]

At its core, diplomacy asserts US power. Modern nation-state power projection extends beyond the simple construct of the air, land, maritime, cyber, and space domains. Theorist Joseph S. Nye Jr. frames modern manifestations of power in terms of "power-conversion strategies."[167] Nye asserts that hard power, soft power, and the hybrid, smart power, are effectively used when one understands the nature of that power (military, economic), the type of behavior required (coercion, protection), the modality it uses (war, sanctions), the qualities for success (competence, trust), and the resources required (troops, budgets) (Figure 6).[168] Nye's depiction of relational power is perhaps the most useful construct to blend political and military effects in the Phase Zero environment.

---

and general guiding principles that are used for military planning.
165 Embassies as Command Posts in the Anti-Terror Campaign, 18.
166 Author interview A03 with Special Forces officer July 18, 2012.
167 Nye, 10.
168 Nye, 42. Nye defines smart power as "the combination of hard power of coercion and payment with the soft power of persuasion and attraction," xiii.

Figure 6. Depiction of Power Behavior Spectrum, adapted from Joseph S. Nye, Jr., *The Future of Power*, 21

Nye theorizes that power has three faces. The first face of power is to get others to do what you want, contrary to their preferences.[169] This is generally recognized as coercive power, often generated from hard power capabilities such as militaries. The second face of power involves the control of framing and agenda-setting.[170] This type of power is less direct in application and favors those who set the forum, framework, and agenda. The United Nations Security Council, with its five permanent members and ten rotational members, is an example of exerting agenda-setting power by the five permanently-seated nations. The third form of power involves influencing what others want: shaping initial preferences and strategies. This type of power involves hidden or subtle influences that channel others' wants, thereby avoiding the need to use coercive or agenda-setting power.[171] This form of power is typified by television marketing campaigns that aggressively promote a product to a certain demographic, thereby shaping desires prior to the contemplation of a purchase. These three forms of power offer insights into how diplomatic strategies might take form. Similarly, these faces of power have corollaries in the special operations power options available in Phase Zero environments.

---

169 Nye, 11. Nye credits Yale political scientist Robert Dahl with defining this face of power in the 1950s.
170 Nye, 12. Nye credits this concept of power to political scientists Peter Bachrach and Morton Baratz.
171 Nye, 13. Nye credits sociologist Steven Lukes with defining this third face of power.

Phase Zero: Contemporary Application

Phase Zero operations, actions, and activities are defined by creative, bottom-up innovations and deliberate, top-down processes. The 2006 and 2010 National Security Strategies and the Global War on Terror (2001-2009) were the significant drivers for the Departments of State and Defense and USAID to more effectively synchronize diplomacy, defense, and development (3Ds). Cognizant of their diverging organizational cultures, peculiar professional languages, and interoperability challenges, the three agencies have forged improved cooperation through collaborative forums, regional programs, and country-specific synchronization. A decade (2001-2011) of interoperability and synchronization lessons are showing through in professional discourse and practical guides. Such works include the *Civil Military Operations Guide*, the *Interagency Conflict Assessment Framework*, and the *3D Planning Guide: Diplomacy, Development and Defense*.[172] These type of tools attempt to "build understanding and synchronize plans to improve collaboration, coordination and unity of effort to achieve the coherence needed to preserve and advance US national interests."[173] These initiatives are formalizing what has been routinely practiced by astute Phase Zero practitioners: translating, understanding, and integrating the programs, processes, and goals of other agencies into one's own plans and programs to achieve a synchronized effect.

---

172 US Agency for International Development, "Civil Military Operations Guide," Verson 2.3 (Draft) February 2012.
US Department of State, *Interagency Conflict Assessment Framework*, July 2008, Department of State website. http://www.state.gov/documents/organization/187786.pdf (accessed January 14, 2013).
3D Planning Guide: Diplomacy, Defense, Development, 15 AUG 2011, Pre-decisional working copy in draft led by the US Agency for International Development. This draft may be superceded by the "Civil Military Operations Guide."
173 3D Planning Guide, 4.

The Department of Defense Phase Zero Framework

The Department of Defense Phase Zero construct is the Theater Campaign Plan (TCP). Theater and Global Campaign Plans are the centerpiece of DoD's planning construct. They provide "the means to translate combatant command theater or functional strategies into executable plans."[174] Combatant Commands design Theater Campaign Plans to orient, integrate, and prioritize resource allocation and operational focus in their region.

Whereas the Theater Campaign Plans are regional (combatant command) synchronizing tools, the Department of Defense concurrently employs a Global Campaign Plan construct. Global Campaign Plans coordinate security issues that exceed the authority or capability of a single combatant commander. Global Campaign Plans are currently active in the realms of cyber, weapons of mass destruction (WMD), and terrorism.[175] USSOCOM is the lead combatant command for combatting global terror networks.[176] Central to the execution of Global Campaign Plans is the ability to see and act across regional boundaries in order to understand, interdict, or disrupt globally agile threats.[177]

The art and science of implementing former Secretary Rumsfeld's notion of improved military engagement in concert with other forms of national power has been labeled as "operationalizing" theater security cooperation.[178] *Security cooperation* is an umbrella term to describe nearly all military

---

174 JP 5-0, xvii.
175 JP 5-0, II-25.
176 Admiral William H. McRaven, USSOCOM Posture Statement before the 112th Congress, Senate Armed Services Committee, March 6, 2012, 22 pages.
177 JP 5-0, II-25.
178 Charles F. Wald, "New Thinking at USEUCOM: The Phase Zero Campaign," Joint Forces Quarterly 43, 4th quarter 2006, 72-73.

activities undertaken in Phase Zero. Security cooperation is defined as:

> activities undertaken by DoD to encourage and enable partners to work with the United States to achieve strategic objectives. It includes all DoD interaction with foreign defense and security establishments, including all DoD-administered security assistance programs, that: build defense and security relationships that promote specific US security interests, including all international armaments cooperation activities and security assistance activities; develop allied and friendly military capabilities for self-defense and multinational operations; and provide US forces with peacetime and contingency access to host nations.[179]

Theater Campaign Plans, while directive and authoritative, are framework documents communicating the strategic narrative, prioritizing resources, and inspiring long-term, programmatic thinking by the executing services and agencies.[180] The actual operational design of Theater Campaign Plan implementation occurs at lower echelons. This process decentralizes the tactical and operational design to the executing service component commands, joint task forces, joint interagency task forces, or sub-unified commands. In terms of operational art, the combatant command establishes the strategic objectives, resource allocation, and regional forums as the broad guidance for decentralized, detailed implementing operations.

---

179 Department of Defense Directive (DoDD) 5132.03, 24 OCT 2008.
180 Department of Defense, *Theater Campaign Planning: Planner's Handbook*, February 2012, Version 1.0. Office of the Deputy Assistant Secretary of Defense for Plans, Office of the Under Secretary of Defense for Policy (Washington DC: US Government Printing Office), 1.

For USSOF, the operational-level application of critical and creative thinking to achieve these objectives occurs at the Theater Special Operations Command (TSOC). The Theater Special Operations Command is the central Phase Zero operational art headquarters that then sets the framework (posture, programs, exercises, regional forums, engagements) for the tactical execution required to link tactics to strategy.

The Theater Campaign Plan applies the logic of operational art with one important difference: instead of arranging battles to win a war, the Theater Campaign Plan arranges engagements to sustain the peace. While battles are narrowly defined as two combatants tactically engaged in a lethal fight, engagements have no such parameters. Engagements include nearly all forms of contact *short of* battles: exercises, foreign military sales, joint combined exchange training, professional schooling, port visits, counter-narcotics training, civil projects, international officer exchange education and training programs, conferences, and media events. Authorities and programs used in Phase Zero include Title 10, Title 22, Title 32, Section 1206, 1208, 1210, counter-narcotics, military-to-military programs, and all forms of security assistance (twelve programs total including Foreign Military Financing (FMF), International Military Engagement Training (IMET), and Peacekeeping Operations (PKO).[181] Theater Campaign Plans are "iterative and often cobble together the various types of resources into a coherent, actionable plan."[182]

At the combatant command level, the special operations contribution in Phase Zero is broadly characterized as a *shape* and *deter* function. Special operations joint doctrine states the role of SOF in Phase Zero:

---
181 Department of Defense, *Planner's Handbook*, 16.
182 Department of Defense, *Planner's Handbook*, 2.

Use of SOF [special operations forces] and SO [special operations], concurrent with conventional force [CF] capabilities in military engagement, security cooperation, and deterrence activities help shape the operational environment and keep the day-to-day tensions between nations or groups below the threshold of armed conflict, which serves to maintain US global influence.[183]

While these definitions provide clear guidance, the modern application of special operations Phase Zero operations is less simple. SOF Phase Zero ranges from routine peacetime engagements (Nepal, Romania) to publicly acknowledged but sensitive, intrusive operations (Colombia, Yemen, Philippines). The employment of special operations in foreign environments often contains complicated caveats, blurred distinctions of police and military roles, lawless conflict zones, and politically charged environments. New technologies such as drones have further complicated Phase Zero with intrusive collection and interdiction options with disputed legal parameters. As a result, Phase Zero is not a fixed category; it has shifting boundaries and blurred edges commensurate with its changing environment. The rise of non-state threats and slow-boiling, low-intensity conflicts have raised important questions: what types of US activities occur in Phase Zero?

This growing ambition of what can be accomplished in Phase Zero comes from directed guidance and self-generated aspirations for what can and should be accomplished prior to crisis, or in terminology of SOF leadership, "left of

---

[183] JP 3-05, I-3.

the bang."[184] While military operations can be quickened and expanded quickly, policies often cannot. Thus, special operations activities in Phase Zero that stretch policy parameters can threaten to overwhelm or outpace the policy itself.

Summary

Phase Zero combines all the activities of the US that engage and shape foreign environments with the intent of preventing (and/or mitigating) conflict and advancing US interests. The lead means is the art of diplomacy. Phase Zero approaches are typically non-lethal with the military focusing on threat and friendly assessments, train-and-equip programs, capacity buildings, and military-to-military contacts.

The last decade of engagement and GWOT principles forged a new notion of Phase Zero as an operation unto itself that requires detailed planning, programs, and execution to arrange US diplomatic, defense, and development resources in a synchronized and effective manner. The art of Phase Zero is fusing disparate activities to accomplish the objectives of US diplomacy, defense, and development efforts. A key challenge lies in the details of execution: exactly how military engagement operations, actions, and activities are blended with development and diplomacy to accomplish strategic objectives.

---

184 Sandra I. Erwin, "Special Operations Command Seeks a Bigger Role in Conflict Prevention," National Defense magazine, November 29, 2012, accessed on December 17, 2012 at http://nationaldefensemagazine.org/blog/Lists/Posts/Post.aspx?ID=983>.

3
Phase Zero Tensions

In the 1966 spaghetti western *The Good, The Bad and the Ugly*, director Sergio Leone weaves a tale of collaboration, deception, and the rugged pursuit of riches between the *Good* Blondie (Clint Eastwood), the *Bad* Angel Eyes (Lee Van Cleef), and the *Ugly* Tuco (Eli Wallach).[185] Throughout the movie, the trio cooperates well when faced with external threats but resorts to predatory actions when left alone. *The Good, The Bad and the Ugly* is a story of team collaboration colliding with shrewd self-interests. Ultimately, none of the trio prove to be good, bad, or worse than the other. Each pursues his aims according to his risk-reward calculation and worldview.[186] The Phase Zero environment is similar: competing agendas, entities, and agencies, both US and foreign, pursue their interests. Within this pursuit exists the persistent tension of calculation, posturing, and seemingly small actions with oversized effects.

Chapter three analyzes Phase Zero tensions. Tensions are

---

185 *The Good, The Bad and The Ugly*, directed by Sergio Leone, Metro Goldwyn Meyer, 2003 Special Edition DVD release (originally aired 1966).
186 Walter A. McDougall, *Promised Land, Crusader State* (New York: Houghton Mifflin Company, 1997), 1-3. McDougall begins the book with a metaphorical spin on *The Good, The Bad, and The Ugly* as representative of American politics in the international arena ("idealistic, hypocritical, and just realistic, often at the same time").

first examined through the lens of organizational cultures. Next, analysis is conducted on the challenges of synchronizing the host nation, the US embassy, the combatant command (DoD writ large), and special operations (within DoD). This synchronization process is analyzed in three sections: policy, programs, and posture. The policy section, the focal point of chapter three, contains vignettes on special operations in Yemen, Indonesia, and Thailand. Chapter three concludes with synchronization observations and insights.

Section I. The Sources of Phase Zero Tensions

Phase Zero challenges all US diplomatic, development, and defense efforts to synchronize and achieve commonly understood and accepted goals. When harmonized, these US government efforts, coupled with commercial US economic, social, and informational influences, offer compelling prospects for a host nation. In theory this is simple. In practice, it is complex, precarious, and rife with tensions.

The Marshall Plan in post-WWII Germany provides a supreme example of smart, harmonized US power. Initiated in April 1948, the Marshall Plan is credited with the rapid economic recovery of war-ravaged European states, the promotion of free and fair markets managed by democratic governments, and the integration of Europe as a single economic community. The Marshall Plan fostered prosperity and nurtured a system that federated European interests, a prescient hedge against future continental wars.[187] In a post-war reconstruction environment, the Marshall Plan effectively unified US diplomatic might, development assistance, and military

---
187 Robert Payne, *The Marshall Story: A Biography of General George C. Marshall*, (New York, New York: Prentice Hall), 1951, 303-321.

power, binded together with a democratic ideology. Today's Phase Zero strategies aim for this exact effect. A critical difference is that today's Phase Zero is a more modest and decentralized proposition that occurs within sovereign countries often challenged by internal instability and "rhizomatic" threats.[188]

The obstacle to harmony in Phase Zero is the tension between the perspectives, practices, and cultures of the US diplomatic, development, and defense communities. Where SOF are present, an additional element of friction may be added. The reason may be organizational more than cultural or cognitive. Simply, special operations perspectives may not be captured by military country team defense representatives such as defense attaches (intelligence-focused) or security assistance officers (program-focused).

Phase Zero tensions arise when the actions of US agencies merge within the narrow confines of a host nation. In this environment, a logical and reasonable act conducted by one agency might be entirely counterproductive for another. For example, visible military actions (DoD) improve security but strain diplomacy (DoS).[189] US military humanitarian aid delivery can achieve clear objectives yet can undermine the perception of a host nation government's ability to support its population in a crisis. When this occurs, diplomatic relations become strained. Similarly, diplomatic decisions may increase

---
[188] Max Manwaring, "Ambassador Stephen Krasner's Orienting Principle for Foreign Policy (and Military Management) – Responsible Sovereignty," (Carlisle Barracks, PA: US Army War College) Monograph, Strategic Studies Institute, April 2012. In his writings on contemporary security threats, Manwaring discusses the rise of "rhizomatic" threats that have "an apparently hierarchical system above ground- visible in the operational and political arenas, and with another system centered in the roots underground" 29.

[189] The US raid on May 2, 2011 in Abbottabad, Pakistan to capture or kill Osama Bin Laden is an example of the tension between security actions and the diplomatic aftermath. Whether the raid was successful or not, diplomatic relations with Pakistan were clear to be damaged.

strategic leverage but raise new security risks by narrowing tactical military options. The US military advisory effort in Mali was diplomatically restricted from accessing the dangerous northern Mali region.[190] While a diplomatically prudent measure, the US had poor intelligence and few options when Al Qaeda in the Islamic Mahgreb (AQIM) occupied northern Mali (Gao, Timbuktu) in 2012.[191]

Development projects, despite efforts to be non-political, also contribute to tensions. Development projects, however altruistic, are partisan acts with social consequences that, if not implemented properly, may resonate in unknown (even harmful) effects well after project completion.[192] Each discipline (defense, diplomacy, development), no matter how expertly executed, is bound to have a splinter effect. These US interagency tensions magnify when interacting with a host nation that has its own competing agendas and aspirations on how to benefit from US largesse.

The diverging perspectives of diplomacy, development, and defense officials can be better understood by examining organizational norms and cultures. Official representatives can be expected to interact with agency-centric viewpoints and unique proclivities.[193] These seemingly superficial differences often furnish the tensions that arise. At the risk of oversimplification, the perspectives of each are juxtaposed below.

Diplomacy places great emphasis on continual

---

190 Author interview A07 and A21 with Special Operations Forces officers directly involved in SOF engagement in Mali.
191 Author interview A07 and A21.
Craig Whitlock, "U.S. counterterrorism efforts in Africa defined by a decade of missteps," Washington Post, February 4, 2013, http://www.washingtonpost.com/craig-whitlock/2011/02/28/AB5dpFP_page.html, (accessed February 9, 2013).
192 Author interview A06 with USAID representative Dale Skoric, August 13, 2012.
193 Anthony H. Cordesman, "Department of Defense, State Department, USAID and NSC Reporting on the Afghan War," May 19, 2010, Center for Strategic Studies and International Studies (CSIS), http://csis.org/publication/department-defense-state-department-usaid-and-nsc-reporting-afghan-war (accessed January 14, 2013).

relations and negotiations akin to astute business practices.[194] Diplomatic protocols are paramount, interpersonal skills are valued, and erudition and strong field credentials are markers of respect. Respect and honor quotients within host nations are critical for effective diplomatic relations and establishing important trust relationships. The State Department values precision in writing and speaking over planning and briefing.[195]

Development cultures are a mix of altruistic energy and programmatic realities.[196] USAID, as the principal agent of foreign non-military assistance, conducts nearly all development projects through contractors, adding another layer of uncertainty in the potentially profitable realm of development work. Development officials are also cautious of association with military actions. For sound reasons, development work too closely associated with military action can invite unwanted risks. They may portray development actions in a martial light or confuse host nation populations who may associate uniformed personnel with humanitarian aid and not long-term development. Development teams value budgets, monitoring, and evaluation of impacts.[197]

Military cultures view the world through the lens of security threats and vulnerabilities. Military personnel are

---

194 Rajiv Chandrasekaran, *Little America: The war within the war for Afghanistan* (New York: Random House, Inc., 2012), 171-189. In his critique of US actions in Afghanistan, Chandrasekaran criticizes the US State Department culture as "observe-and-report." Chandrasekaran's past decade of reporting offers a detailed description and narrative about US civilian agencies in Afghanistan (and Iraq).

195 US Department of State, official website, "13 Dimensions of a Foreign Service Officer," http://careers.state.gov/resources/downloads/downloads/13-dimensions, (accessed January 15, 2013). The comment that writing and speaking is valued over planning and briefing is not the official position of the US State Department. This is the observation of the author in working with both military and State Department personnel.

196 Tiaji Salaam-Blyther, "USAID Global Health Programs FY2001-FY2012 Request," Congressional Research Service, June 330, 2011, 10-11.

197 Author interview, Dale Skoric, August 13, 2012.

action-oriented with infectious, if overbearing, can-do attitudes. With huge capacity and a myriad of programs across all armed services, military actions can be overwhelming, military officials can be myopic, and the totality of military programs in a country often lacks a single synchronizing official. Armed services do bring assertive leadership, collaborative skills, broad professional competencies, and credible capacity to accomplish difficult goals in challenging environments. Military-to-military relationships are also easily established with long-term relationship and capability benefits for all sides. Military cultures value planning.[198]

Special operations representatives, a subculture in the defense community, bring a dual-natured perspective. The indirect perspective favors the delivery of low footprint, highly skilled assistance to host nation security forces. Direct SOF approaches specialize in delivering precision, lethal effects against human and network targets through unilateral or partnered methods. SOF are relationship-centric. Unlike services that employ large platforms (ships, planes), special operations are heavily human-oriented. SOF bring regional expertise through basic language proficiency and modest cultural exposure. While cultural acumen helps breed trust, SOF are also associated with clandestine methods and human targeting programs, a potential antibody to developing trust. SOF are typically older with an expectation of great autonomy.

---

[198] Andrea Barbara Baumann, "Clash of Organizational Cultures? The Challenge of Integrating Civil and Military Efforts in Stabilisation Operations," RUSI Journal 153, No. 6, December 2008, 70-73, http://www.rusi.org/downloads/assets/Baumann.pdf (accessed January 14, 2013). Baumann focuses on the deeper philosophical disagreements between the military and civilians conducting stabilization operations. The cultural descriptions of each agency are the author's and not Baumann's.

PHASE ZERO TENSIONS

Photo 6. Civil affairs teams apply unique capabilities to assist remote populations with livelihood and health issues (US Army photo, authorized use).

The cultural and organizational tensions inherent in these agencies are certainly reconcilable. Relationships of trust built on credibility, communication, and professional competency will not erase these tensions, but can nullify their corrosive effects. Bridging these cultural divides requires sophistication, patience, and a reasonable willingness to identify and mitigate these tensions. It also requires detailed frameworks and explicit methods that connect the policy logic to the commitment of US resources.

Section II. Synchronization Frameworks

Effective action in Phase Zero involves synchronizing host nation equities, US policy, US-sponsored programs, US military posture, and the people and processes that create unity of effort (Figure 7).

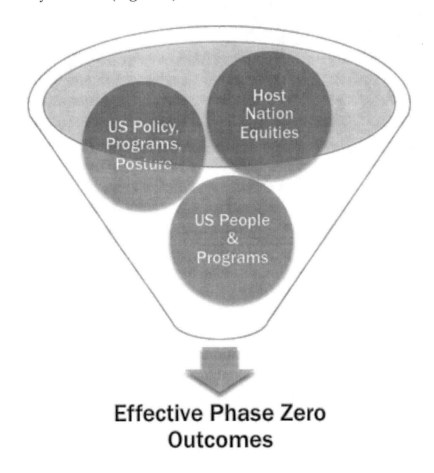

Figure 7. Effective Phase Zero Outcomes

The US views host nation interests through the lens of the host nation's Internal Defense and Development (IDAD) strategy. IDAD is "the full range of measures taken by a nation to promote its growth and to protect itself from subversion, insurgency, lawlessness, terrorism, and other threats to security."[199] In Phase Zero, US policy contributes to and shapes a host nation's IDAD strategy. Most host nations where the US contributes Phase Zero assistance are contending with some form of instability, civil strife, governing weaknesses, or external threats. An IDAD strategy can be clear and codified (Plan Colombia) or entirely lacking in coherency (Somalia). The aim of the US is to support, but not supplant, a host nation's internal plan. Overwhelming US assistance can be counterproductive if the host nation is viewed as a mere recipient of US assistance on terms dictated by the US. To properly support a host nation strategy, US diplomacy, development, and defense agencies conduct strategies guided by their staffed and approved execution plans. Each is discussed below.

The US Embassy country team, led by DoS, is guided by their Country Campaign Plan (CCP) and Integrated Country Strategy (ICS).[200] The ICS is a "country-level strategy, budget allocation and request, and performance assessment document."[201] The country-specific plans are informed and guided by the Department of State Joint Regional Strategy (JRS) that outlines the most significant regional foreign policy goals and resource requirements.[202]

Since 2004, the Department of State and USAID have issued a joint strategic plan. From these strategic priorities, USAID

---

[199] JP 1-02, 55. IDAD is a US term, not a host nation term.
[200] The Integrated Country Strategy (ICS) replaced the Mission Strategic Resource Plan (MSRP) in 2012.
[201] 3D Planning Guide, 5.
[202] The US Department of State divides the world into six regional bureaus under the Underscretary for Political Affairs. The six bureaus are African Affairs (AF), East Asia and Pacific Affairs (EAP), Europe and Eurasia Affairs (EUR), Near Eastern Affairs (NEA), South and Central Asian Affairs (SCA) and Western Hemisphere Affairs (WHA). The officer of the Undersecretary for Political Affairs is the "day-to-day manager of the overall regional and bilateral policy issues." US Department of State website, accessed on December 18, 2012 at http://www.state.gov/r/pa/ei/rls/dos/436.htm
The Joint Regional Strategy replaced the Bureau Region Strategic Plan (BSRP) in 2012.

develops a Country Development Cooperation Strategy (CDCS).[203] At the lower, execution level, USAID uses an Operational Plan (OP) that allocates resources and directs the means of implementation. To look regionally, USAID assembles (or joins) joint planning cells comprised of Department of State, US Department of Agriculture (USDA) and others including the US military. When employed, these joint planning cells are powerful mechanisms for synchronization.[204]

The US military takes Phase Zero theater-strategic guidance from the combatant commander issued through the Theater Campaign Plan (TCP) and Global Campaign Plans (GCP). Though not standardized, the US military generally devises a Country Action Plan (CAP) that logically connects the combatant command efforts with the country team Integrated Country Strategy. The Country Action Plan serves as a concise (5-10 pages) road map of how, where, and why US military capacity is applied within a country.

Country Action Plans use "lines of effort," a joint doctrinal term that describes the purpose of an activity (non-geographic) as a categorical descriptor of related military intentions and programs.[205] Example Country Action Plan lines of effort include build host nation military capacity, develop host nation institutional capability, improve information sharing, and improve port maritime control. Country Action Plans aim to reflect, on a very broad level, the totality of US military efforts within each country. Capturing the

---

203 Civil-Military Operations Guide, draft version 2.3, February 2012. This guide, still in development, was developed by USAID's Office of Civil-Military Cooperation. The draft guide is referenced here because it contains the latest and most accurate descriptions of "3D" documents, processes and organizations.
204 Author interview A06 with USAID representative, August 13, 2012.
205 JP 1-02. Line of Effort (LOE): "In the context of joint operation planning, using the purpose (cause and effect) to focus efforts toward establishing operational and strategic conditions by linking multiple tasks and missions," 185.

sum total of all military efforts can be a challenge, particularly when all five armed services are applying significant, persistent engagement underpinned by major arms sales or procurement programs.

Special operations Phase Zero contributions fall into four categories, none of which are neatly defined by a single codified document. The categories are: situational awareness/understanding; craft and expand access venues; develop and nurture human and physical networks; and provide capability and capacity building to host nation security forces.[206] The majority of special operations efforts are typically reflected in the Country Actions Plans. In certain cases, Country Action Plans are written by special operations representatives, particularly when USSOF is the lead or majority military engagement effort in the country.[207]

These framework strategies and documents provide the cognitive direction and programmatic orientation of each US agency and the host nation. While not hierarchical in nature, the country team is the *ipso facto* synchronization lead. The US Ambassador, by extension and by the authority of a presidential appointment, exercises command-like authority within a country for all US government sponsored actions. With few exceptions and within legal parameters, if the US Ambassador directs a start, change, or stop to an operation, action, or activity, then it is his or her prerogative to do so. A SOF officer with US embassy experience captures the nuances of this relationship:

---

206 Author interview A08 with Special Forces officer, 06 August 2012.
207 At the inception of the Trans-Sahel Counterterrorism Initiative (TSCTI) in European Command (2005-2007 period), Country Action Plan development was led by Special Operations Command Europe (SOCEUR) in conjunction with COCOM guidance. Similarly, Special Operations Command Pacific (SOCPAC) took the lead in the mid-2000s in developing similar products for countries with growing SOF engagements.

It is not quite accurate for DoD to complain that we have the authorities to do anything but the approvals to do nothing ... SOF can be very powerful agents of good interagency cooperation and expertise but only once you've built trust and relationships with members of the team, you are interacting socially, you are demonstrating competence, doing the small things well ... this is not cloak and dagger stuff or derring-do, this is the Ambassador saying 'when I talk to this officer, he's knowledgeable, he doesn't try to speak outside of his lane, he doesn't fake any expertise and he understands my vision for the country and where I'm comfortable and potentially uncomfortable with DoD.'[208]

Even with a clear understanding of the Ambassador's role and coherent organizational strategies, the implementing details of engagements prove difficult to arrange in time, space, and purpose. The outcomes of these strategies are policy interpretation, program implementation, and armed forces posture. These efforts are thoroughly human endeavors and not a mechanistic process. That is what makes them so difficult. Policy, programs, and posture are analyzed below to illustrate the complexity of frictions in Phase Zero.

Section III. Synchronization Challenge One: Policy

The implementing activities in Phase Zero are an attempt to "translate the strategic and policy guidance to actions that get the job done on the ground."[209] Deconflicting and syn-

---

[208] Author interview with US Special Forces officer, LTC Josh Walker, 06 August 2012.
[209] Author interview A05 with COL (Ret.) David Maxwell, Associate Director, Center for Security Studies and Security Studies Program, Edmund A. Walsh School of Foreign Service, Georgetown University, July 30, 2012.

chronizing the myriad of activities aiming to achieve policy goals is one challenge of implementing policy. A second and more difficult challenge is the widely interpreted and even disputed actions that should be applied against and within a policy. This second policy challenge is often a concern with special operations that are conducted within and across the borders of a sovereign host nation.

To illustrate policy challenges germane to special operations in Phase Zero, three Phase Zero policy vignettes are examined: an *incremental* example of counterterrorism campaigning in Yemen; a *legislative* example of the 1997 Leahy Amendment requiring human rights vetting prior to US military assistance; and a *political* example of sustained US engagement in Thailand despite a historical propensity for non-democratic regime change (coup d'etats).

Incremental Policy Vignette: Yemen

In September 2011, then-CIA Director David Petraeus proclaimed the Yemen-based Al Qaeda in the Arabian Pennisula (AQAP) "the most dangerous regional node in the global jihad."[210] Based on foiled plots to attack the US in 2009 (US bound airliner on Christmas day) and 2010 (parcel shipments), AQAP asserted itself as a regional jihad group with the intent and capability to attack the US homeland. On October 04, 2012, the US State Department declared the Yemen-based Ansar al-Sharia (an AQ affiliate) as a Foreign Terrorist Organization (FTO).[211]

---

210 Reuters, "CIA Chief: Yemen Qaeda most dangerous," Reuters online news service, September 13, 2011, http://www.reuters.com/article/2011/09/13/us-usa-security-qaeda-idUSTRE78C3G720110913 (accessed January 15, 2013).
211 Congressional Research Service, "Yemen: Background and US Relations," Jeremy M. Sharp, November 12, 2012, 12. http://www.fas.org/sgp/crs/mideast/RL34170.pdf (accessed December 15, 2012).

To combat AQAP and Ansar al-Sharia in their Yemeni safe havens, the US increased its aid to Yemen from $61.9 million in 2006 to $316.4 million in 2012.[212] In early 2010 the US authorized $155.3 million in security assistance to Yemen with $34.5 million appropriated for Yemeni special operations forces counterterrorism operations.[213]

Providing assistance to Yemen (Figure 8) remains complicated by five factors: a Yemeni civil war; the pervasive influence of Yemeni security forces by former President Ali Abdullah Saleh; a controversial legal footing for US-sponsored targeting operations; a Yemeni population sensitive to US influence;[214] and the involvement of US citizens (Anwar al-Awlaki, killed September 30, 2011 by a drone strike) in Yemeni-based AQAP organizations.[215]

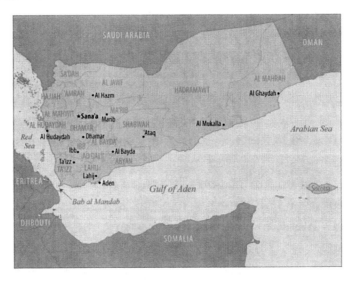

Figure 8. Map of Yemen, CRS, July 2012

212 CRS report, 20
213 CRS report, 20.
214 2011 Yemen Stability Survey, March 2011, Glevum Associates, LLC. In a survey of 1005 Yemeni adults, only 4% surveyed "somewhat or strongly approve" of Yemeni government cooperation with the US.
215 CRS report.

Given this context and the sensitivities involved, US policy and in-country actions in Yemen are built and applied incrementally. According to a Congressional Services Report:

> U.S. aid to Yemen in any given fiscal year can come from as many as 17 different aid programs managed by multiple agencies, including the Department of State, USAID, and the Department of Defense. Annual appropriations legislation specifically requires the executive branch to notify the Appropriations Committees prior to the obligation of funds for programs in Yemen.

The operational art challenge is the management of seventeen programs – each with its own programatic timelines - to achieve a synchronized effort. These often-cumbersome programs are designed around bureaucratic orderliness rather than military precision. The result is incremental delivery and application. Even cohesive packages become large administrative matters foisted on small, forward-based military teams. In the National Defense Authorization Act (Section 1207), authorized on June 7, 2012, DoD and DoS jointly certified $75 million in aid to Yemen's Minister of the Interior Counterterror Forces.[216]

> Assistance may include the provision of equipment, supplies, and training, as well as assistance for minor military construction, for the following purpose: "To enhance the ability of the Yemen Ministry of Interior Counter Terrorism Forces to conduct counter-terrorism

---

216 CRS report, 15. "Section 1207 (n) (C) of the FY2012 National Defense Authorization Act (P.L. 112-81) established a new transitional authority that would permit the Secretary of Defense, with the concurrence of the Secretary of State, to assist counterterrorism and peacekeeping efforts in Yemen during FY2012."

operations against al-Qaeda in the Arabian Peninsula and its affiliates"...The GSCF FY2012 aid includes, among other things, funds for night vision goggles, armored wheeled vehicles, and operational training.[217]

This package provides substantial military aid that requires a thoughtful and deliberate training and assistance campaign. Beyond the mere components of these programs, operational art is required to connect the largesse of military aid with the US-Yemeni campaign objectives. The US government assistance to Yemen, both development and military aid, is further complicated with a publically debated question: Is the US at war in Yemen? Or is Yemen what steady-state Phase Zero operations look like in today's threat environment?

Yemen provides a vexing policy case of joint US-Yemeni actions against AQAP networks within a troubled Yemeni political arena. The September 20, 2011 drone strike on US citizen and Ansar al-Sharia member Anwar al-Awlaki raised the public profile of US actions in Yemen. To increase the transparency of US actions in Yemen, on June 15, 2012 US President Barack Obama publically declared that the US is actively engaged in joint (US-Yemeni) hostile actions against AQAP in Yemen and Somalia.

> The U.S. military has also been working closely with the Yemeni government to operationally dismantle and ultimately eliminate the terrorist threat posed by al-Qa'ida in the Arabian Peninsula (AQAP), the most active and dangerous affiliate of al-Qa'ida today. Our joint efforts have resulted in direct action against a limited number of AQAP operatives and senior leaders in

---
217 CRS report, Section 1207, 15.

that country who posed a terrorist threat to the United States and our interests.[218]

The details of operations in Yemen are largely classified. However, the public articulation of US actions in Yemen is increasingly declarative. From the appropriations, programs of record, publically acknowledged operations, and declarations from The White House, there is clearly an interagency, campaign-style approach in Yemen.[219] Examining the relationship between intelligence, development, diplomacy, surgical strike, and special warfare is fundamental to assessing such a special operations (or interagency) operational art. Such an examination in Yemen is beyond the scope of this study.

The incremental nature of US policy in Yemen demonstrates a cautious and responsible escalation of support to Yemen with the full visibility and concurrence of the executive and the legislative branches. It also represents how incremental policies and programs, bound together by small but capable US special operations forces and interagency partners, can expand the Phase Zero paradigm into a construct suitable for creative and effective operational art. This operational art must weave together a campaign from existing authorities; establish discreet but capable command and logistics nodes; and integrate partner-building, development assistance, and human targeting strikes within *a politically sensitive environment*.

---

218 President Barak Obama, "Presidential Letter – 2012 War Powers Resolution 6-month Report" (Washington DC: The White House Office of the Press Secretary), June 15, 2012, http://www.whitehouse.gov/the-press-office/2012/06/15/presidential-letter-2012-war-powers-resolution-6-month-report (accessed December 18, 2012). Until this letter, US actions in Yemen were characterized as unofficially acknowledged by the press and by extension, the US government.
219 JP 5-0, GL-6. A campaign is defined as "A series of related major operations aimed at achieving strategic and operational objectives within a given time and space."

In this charged political environment, the open declaration of a US military campaign, SOF-led or otherwise, would invite great risk with near-certain counterproductive outcomes. An alternative is a nascent campaign, woven from patchwork programs, postures, and agreements. An embryonic campaign – one executed without a declarative moniker such as Operation Enduring Freedom - can be more judiciously calibrated to pursue tactical, operational, and strategic objectives. Yemen and the Arabian Peninsula typify the sensitive environments where these campaigns are differentiated options available to policy makers. Whether publically declared a campaign or not, such special operations applications fit the campaign definition of "a series of related major operations aimed at achieving strategic and operational objectives within a given time and space."[220] Publically acknowledged but discreet actions in locales like Yemen increasingly have the logic, synchronization, and characteristics of an interagency campaign. In order to achieve results at the tactical, strategic, and diplomatic levels, these campaigns require skilled operational and diplomatic art, applied with great finesse and insight by a synchronized interagency team.

The Yemen case shows the art of the possible on crafting a distinctive, other than major military intervention type of Phase Zero campaign. The potential for a special operations-focused Phase Zero operational art has been made possible by incremental changes in US security cooperation programs. "The broad definition of security cooperation to include all DoD international programs and those seven FAA/AECA [Foreign Assistance Act / Arms Export Control Act] authorized programs administered by DSCA [Defense Security Cooperation Agency] has significantly increased the playing

---

[220] JP 1-02, 37.

field for DoD."²²¹ US actions in Yemen seek to balance the requirements of operational secrecy, public transparency, executive writ, and legislative oversight while coping with the regional sensitivities of intrusive US actions in the Arabian Peninsula.

A larger legal, political, and moral debate, beyond the scope of this vignette, is at the heart of discreet US operations in peacetime environments. Importantly, the Yemen case illustrates how US policy, adjusted on the margins and incrementally applied, is sufficient to enable a discreet, scalable, joint SOF campaign approach that is "left of the Joint Task Force threshold."²²² The campaign requirements in locales such as Yemen align well with USSOF attributes and core competencies. The operational design that concocts the organizing logic of such a campaign must blend the primacy of policy with a vast mix of programs implemented, in part, by small USSOF teams.

Legislative Policy Vignette: 1997 Leahy Amendment

The "Leahy Amendment" or "Leahy Act" is shorthand for a 1997 amendment to the Foreign Assistance Act. Named after Vermont Senator Patrick J. Leahy, the "spirit and intent of the act is to force foreign governments to take responsibility for, investigate, and prosecute cases of human rights abuses. Leahy is really not about individuals or units, it is about pressuring the governments."²²³ The amendment stipulates:

---

221 "The Management of Security Cooperation (Greenbook)", Defense Institute of Security Assistance Management (DISAM), February 2012, 31ˢᵗ edition, http://www.disam.dsca.mil/pages/pubs/greenbook.aspx, (accessed December 18, 2012)
222 United States Special Operations Command (USASOC), unclassified command brief, September 2011.
223 Charles "Ken" Comer, "Leahy in Indonesia: Damned if you do (and even if you don't), Asian Affairs: An American Review 37:53-70, 2010, Francis & Taylor Group, LLC, 66.

No assistance (includes both articles and training) authorized by the FAA [Foreign Assistance Act] or the AECA [Arms Export Control Act] will be made available to any unit of the security forces of a country if the Secretary of State has credible information that such unit has committed a gross violation of human rights. Funding may be provided once the secretary determines and reports to Congress that the affected country is taking effective measures to bring the responsible members of the security forces unit to justice [section 620M, FAA] ... Proposed students and/or units are to be vetted using all available USG [United States Government] resources prior to any training or combined exercises.[224]

Since its passage, the Leahy act has blocked military assistance (weapons funding and training) to countries such as Colombia, Indonesia, the Philippines, and Pakistan.[225] In the case of US relations with Indonesia, the amendment represents a policy dilemma.

The Southeast Asian country of Indonesia is the world's fourth most populous nation (238 million) with a predominately Muslim population. Indonesia is a strategically located north of major US ally, Australia, and south of the Philippines and Malaysia (Figure 9). It lies astride the Malacca Straits, the world's fourth largest shipping lane. US strategic interests include economic and commercial trade, security (counterterror), and competition for influence in Southeast Asia.

---

224 DISAM Greenbook, 2-10.
225 June S. Beittel, "Colombia: Background, US Relations, and Congressional Interest," Congressional Research Service, November 28, 2012, 1-26.

Figure 9. Map of Indonesia, CRS, November 2011

The decade of the 2000s witnessed two significant trends in US-Indonesian relations. First, Indonesia showed progress in areas of concern for the US, principally in representative government and security sector reform, including human rights. Second, the US GWOT strategy sought improved visibility and access to Southeast Asia terror threats. The threat that most concerned the US was the presence of Al Qaeda affiliates in the Celebes Sea bordering nations of Indonesia, Malaysia, and the Philippines.[226] As the US and Indonesia attempted to move closer in relations, two obstacles prevented closer collaboration. The first was the enforcement of the Leahy Act, which prohibited the growth of US bilateral military engagement with Indonesian Army (TNI) units.[227] The second obstacle was the negative effects of the US GWOT in Indonesia, plummeting the perceptions of the US and, subsequently, reducing US soft power influence and access.[228]

---

226 Zachary Abuza, *Militant Islam in Southeast Asia: Crucible of Terror*, (Boulder, CO: Lynne Rienner Publishers, 2003).
227 TNI stands for Tentara Nasional Indonesia, translated as Indonesia National Armed Forces.
228 Nye Jr., *The Future of Power*, 22, 55. In explaining soft power, Nye describes how Indonesian positive perceptions of the US drastically fell following the initiation of the

This diplomatic impasse was interrupted by the catastrophic December 26, 2004 Indian Ocean tsunami. From this sudden, massive humanitarian crisis and its attendant global attention, the US and Indonesia rapidly expanded collaboration in the realm of humanitarian assistance. The Indonesian government immediately began accepting US humanitarian assistance provided from US Pacific Command, principally in the northwest Sumatra region centered on Banda Aceh. This humanitarian crisis collaboration thawed US-Indonesia relations and opened up the possibility of improved state relations. Toward this end, Secretary of State Condoleeza Rice "exercised a national waiver provision provided to the FY2005 FOAA [Foreign Operations Appropriations Act] to remove congressional restrictions on foreign militiary financing and lethal defense articles on November 22, 2005 and represented a reestablishment of normalized military relations."[229] This waiver triggered an executive and legislative branch thrust-and-parry over US assistance to Indonesia, centered on the Leahy provisions. While the spigot of US assistance was eventually opened wider, US Pacific Command (PACOM) and its sub-unified command, Special Operations Command, Pacific (SOCPAC), were stifled in reopening engagement with units with spotty human right records, most notably the Indonesia Special Forces Command, KOPASUS.[230]

With military engagement to KOPASUS potentially feasible, the US and allied nations aimed to increase collaboration with Indonesia on identifying and defeating radical Islamic terror threats. As part of a larger approach to the whole of Southeast Asia, the US Departments of State and Defense

---

2001 GWOT. Similarly, positive perceptions rose following the US' 2005 humanitarian assistance to victims of the December 26, 2004 Indian Ocean tsunami near Aceh.
229   Comer, Leahy in Indonesia, 60.
230   KOPASUS is an acronym for Komando Pasukan Khusus, or "Special Forces Command."

jointly sought increased engagement with Indonesian counterterrorism police and military units. US law enforcement agencies, in conjunction with Australian Federal Police (AFP), built strong relations with "Detachment 88," Indonesia's police counterterror unit. Many cite the police actions of Detachment 88 against Indonesia-based terror networks as the model of skilled, discreet counterterrorism.[231]

Within the special operations realm, engagement with the counterterrorism units in KOPASUS were stifled. Strategically, US counterterrorism efforts in support of Malaysia and the Philippines were already challenged by the vast, unchecked transit region that included the Indonesian archipelago (Figure 10).

Figure 10. Map of Celebes Sea Region, USGS 2011

---

231 Bruce Vaughn, "Indonesia: Domestic Politics, Strategic Dynamics, and US Interests," Congressional Research Service, October 27, 2010.

The main safe havens for Southeast Asia terrorist threats included the sea and island border region between east Malaysia, the southern Philippines and northern Indonesia.[232] Further complicating this challenge was the inability to grow military relationships, interoperability, and capacity with KOPASUS. A Pacific-based special forces officer stated, "Indonesia was a constant point of frustration. The vetting requirements shut down all engagement, despite clear openings as early as 2006. SOF engagement is about trust models, developing and expanding relationships ... it takes years to develop trust models with true leverage points. We just couldn't get started."[233]

The policy dilemma of the KOPASUS case required measuring the risks and rewards of enforcing principled US legislative actions that prohibits engaging foreign security units deemed important to long-term national interests. The wide interpretations and uneven enforcement of the Leahy Act color this debate. Critics of the act cite waivers gained for engagements in Colombia[234] and Pakistan, but stringent enforcement in Indonesia despite clear improvements in human rights actions.[235]

> It is doubtful that the Leahy standards applied to Indonesia are applied anywhere else in the world, with the notable exceptions of Colombia and Sri Lanka, and they place Indonesia into a category normally reserved for Iraq or Afghanistan. No other nation in the Pacific Rim (except Sri Lanka) must face such scrutiny.

---

232 Abuza, *Miltant Islam*, 121-178.
233 Interview with US Army Special Forces officer, Lieutenant Colonel Adrian Donahoe, August 20, 2012.
234 President William J. Clinton signed a waiver on August 22, 2000 to expand military assistance to Colombia, principally to support counter-drug operations.
235 Comer, Leahy in Indonesia, 62.

Moreover, these types of HRV [Human Rights Vetting] standards are not applied to the two nations that consume roughly 70 percent of all US security assistance granted worldwide – Israel and Egypt.[236]

The KOPASUS vignette demonstrates the tense intersection of policy realities and a putative operational art. PACOM and SOCPAC aspirations to shrink the maneuver and transit space between Malaysia, the Philippines, and Indonesia involved linking the tactics and programs of three countries into a loosely organized regional strategy. This strategy was conceptualized as an iterative approach.[237] This method involved connecting US bilateral relations with the formal and informal cooperative venues already established between Malaysia, the Philippines, and Indonesia. Even without restrictions, the tenuous relations of these countries, vast ocean spaces, and interoperability challenges made this difficult to plan, implement, and synchronize. The inability of SOCPAC to engage a critical partner in this campaign-like approach further inhibited the establishment of a comprehensive regional approach. The combination of these factors stymied the formation of a regional operational art.

To be fair, lamenting over policy restrictions that stifle operational art risks elevating military efficacy over policy primacy. The Leahy Act, despite its controversies and uneven enforcement, is a policy with clear and broad aims regarding US assistance to potential partners. US military engagement is a tool to support these policy aims. The decade-long postponement of USSOF engagement with KOPASUS did

---
236 Comer, Leahy in Indonesia, 62.
237 Author experience at the Special Operations Command – Pacific, June 2006 to June 2008.

generate pressure on the Indonesian government to institute reforms. Thus, the SOF operational art implementation suffered but for the larger and more comprehensive US aspiration of human rights reforms and improved civil-military relations. This vignette illustrates the clear conflict and resulting tensions when policy prohibitions prevent the execution of well-concieved military campaign constructs.

Political Policy Vignette:
Special Operations Engagement in Thailand

In the past 70 years, the US and the Kingdom of Thailand (Figure 11) have sustained consistently productive and friendly relations with mutual benefits in economic, commercial, military, and tourism interests. This consistency is remarkable given the staggering history of non-democratic regime change and political upheaval in Thailand. Since 1932, Thailand has experienced 18 coup d'etats.[238] Underlying this statistic is an implicit truth about foreign relations: friends matter and strategic friends matter more. Thailand, a major non-NATO ally and historically reliable partner vis-a-vis US foreign policy objectives, is a strategic anchor point for the US in Asia.[239] Beyond its influence in Southeast Asia, Thailand has tremendous geo-strategic access to the whole of the Asia-Pacific basin.[240]

---

238 "Counting Thailand's Coups," Nicholas Farrelly, March 08, 2011 on blogsite "New Mandala," accessed on December 21, 2012 at http://asiapacific.anu.edu.au/new-mandala/2011/03/08/counting-thailands-coups/. There is some debate as to what qualifies as a coup d'etat or an attempted coup d'etat. The number is generally between 17 and 18. Wikipedia cites that since 1932, Thailand has had "17 constitutions and charters." Accessed on December 12, 2012 at http://en.wikipedia.org/wiki/Thailand.
239 Emma Chanlett-Avery and Ben Dolven, "Thailand: Background and US Relations," Congressional Research Service, June 5, 2012.
240 The Kingdom of Thailand provided support for the US during the Vietnam war and more recently, supported the 2003 US invasion of Iraq with political backing and military force contributions. Geo-strategically, Thailand sits at the crossroads of Asia with

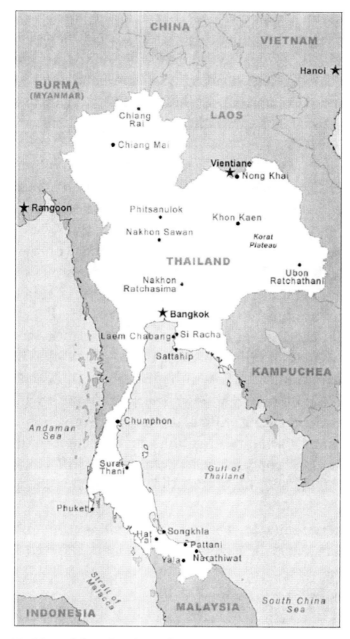

Figure 11. Map of the Kingdom of Thailand, CRS, June 2012

close proximity to South Asia, Northeast Asia, and Southeast Asia.

What then classifies US-Thai engagement as a *political policy vignette*? Amidst the cyclical upheavals in Thai governments, US-Thailand military-to-military relationships have buttressed and steadied US-Thai relations. The historically productive and consistent US military engagement with Thailand is a key facet of maintaining US relations and interests in Southeast Asia.

Thailand periodically teeters below US stated ideals for democratic governance, media repression, and other human rights issues.[241] In 2006, the Royal Thai Army (RTA) staged a coup d'etat that deposed the caretaker government of Prime Minister Thaksin Shinawatra. Despite this non-democratic action, US diplomatic language was relatively muted. Military-to-military relations and programs experienced short disruptions but were maintained.[242] The US response can be interpreted in two ways. The US can be viewed as showing prescient diplomatic patience while allowing Thai politics the time to work through upheaval. A second interpretation might question stalwart US support when Thai political actions undermine the clear US position against coup d'etats. Both interpretations have elements of truth. In 2007, Thailand conducted elections and returned to civilian-control, with few disruptions to US programs, operations, and activities.[243]

Underpinning US-Thai military-to-military relations is a fifty-year history of bilateral engagements between US and

---

[241] Human Rights Watch, "World Report 2011: Thailand, Accessed on December 21, 2012 at http://www.hrw.org/world-report-2011/thailand.
US Department of State, "2010 Human Rights Report: Thailand," Report, 2010 Country Reports on Human Rights Practices, April 08, 2011. Accessed on December 21, 2012 at http://www.state.gov/j/drl/rls/hrrpt/2010/eap/154403.htm

[242] The Nation, "US cuts off millions in military aid to Thailand," September 29, 2006, http://nationmultimedia.com/2006/09/29/headlines/headlines_30014950.php, (accessed December 21, 2012).

[243] Chanlett-Avery and Dolven, CRS report, 12.

Thai special operations forces.[244] In 2011 alone, USSOF conducted over seventeen engagement events in a near-persistent pattern of US-Thai SOF exchanges, training, and exercises.[245] With this persistency in engagement, US-Thai SOF relations are conducted with the depth and trust gained from consistent operational and institutional collaboration.[246]

The focal point for USSOF engagement with Thailand is the Royal Thai Army Special Warfare Command (RTA SWCOM). This command is roughly the Thai equivalent of the US Army Special Operations Command (USASOC).[247] With the Royal Thai Army Special Warfare Command as the main Thai SOF synchronizing headquarters, USSOF engagement has increasingly taken on more predictability and increased scope across the various SOF disciplines (counterterrorism, counterinsurgency, military information support operations).

US military engagement in Thailand is coordinated and synchronized by the Joint US Military Advisory Group (JUSMAGTHAI). JUSMAGTHAI is both a security assistance organization and the office of primary responsibility for all US military bilateral engagement with Thailand.[248] Within JUSMAGTHAI, USSOF officers and noncommissioned officers craft USSOF engagement venues within the overall US engagement strategy. Within the JUSMAGTHAI model, strong USSOF-Thai relations create venues to grow broader US-Thai

---

244 J. "Lumpy" Lumbaca, "The US-Thailand ARSOF Relationship," Special Warfare, Volume 25, issue 1, January-March 2012 (Fort Bragg, NC), 52.
245 Lumbaca, 52. Numerically counting engagements (exercises, training, exchanges) can be a misleading method to characterize engagement without the context of each event. For example, a six week Joint Combined Exchange for Training (JCET) is considered one event as well as a two-person, two-day exchange at a conference. In general, major events are counted as singular engagements with another layer of steady-state engagements (pre-deployment site surveys, informal exchanges, etc) occuring regularly.
246 Lumbaca.
247 Lumbaca, 52.
248 Website of the Joint United States Military Advisory Group, Thailand. www.jusmagthai.com. (accessed on January 4, 2013).

Photo 7. A US Army Special Forces soldier conducts small unit tactics training with partnered forces (US Army photo, authorized use).

relations in the areas of International Military Exchange Training (IMET), the Combatting Terrorism Fellowship Program (CTFP), and multiple US-based education courses.[249] Strategically, US-Thai special operations engagements contribute to SOF interoperability, shared understanding, and regional partnership with deep trust models that survive political upheavals. USSOF engagement in Thailand also strengthens the broader military-to-military relationships that bolster the diplomatic connectivity between the nations.

What are the unique aspects of operational art in the USSOF campaign of engagement in Thailand? This type of operational art is not campaigning in the combined arms sense of joint actions taken against a foe. USSOF engagement in Thailand seeks to create a persistent arc of engagement that is progressive, iterative, and builds on past training and education

---

249 Lumbaca, 53.

venues. When sustained engagements result in institutions like RTA SWCOM, then future engagements move beyond tactical exchanges into more sophisticated arenas. Such areas include military education, security philosophies, specialized training, regional threat assessment, and tailored operational approaches for both internal and external security threats.[250] The RTA SWCOM example illustrates how sustained tactical and relational trust can grow vague cooperation notions into more lasting strategic institutions and actions.[251]

The USSOF engagement model aims to strike a balance between US national security interests and Thai aspirations for improved SOF capability. In countries where coup d'etats are frequent, host nation SOF capacity can be highly political. Cognizant of this dynamic, USSOF in Thailand persistently pursue US objectives in Southeast Asia in conjunction with partnered Thai SOF units. USSOF engagement, while near persistent, is not viewed as overbearing, intrusive, or suspect. Based on long-term collaboration and quality trust models, USSOF engagement hovers at the correct "saturation" level for US resource allocation, Thai political sensitivities, and continuity of programs. This requires a creative application of art and science to build forward progress by weaving together episodic events, programs, and activities.[252] This model is bolstered by carefully emplaced regionally aligned, culturally educated, and Thai-language trained USSOF representatives. These specialists synchronize the tactical engagements and Thai institutional building projects in conjunction with the

---

250 Author experience at Special Operations Command Pacific, 2006-2008.
251 The development of the NATO Special Operations Headquarters in Stuttgart, Germany (now located in Mons, Belgium) is a parallel case of expanding tactical relationships into stronger institutional associations.
252 Author experience at Special Operations Command Pacific, 2006-2008, and 1st Special Forces Group (Airborne), 2008-2010.

JUSMAGTHAI offices.[253] The placement of such SOF representatives in Thailand also has a strategic dividend of providing US insight on Thai perspectives on security. In-country SOF teams help assess how US actions resonate within Thai governmental and population spheres. Additionally, such representatives articulate USSOF programs in order to decrease the frictions within intersecting US programs. A SOF embassy representative explains this role:

> [SOF in-country] representatives act as an interpreter for SOF actions and almost as a lubricant when SOF are on the seam of intelligence and operations. They are well-informed on political sensitivities and can spell out the purpose of SOF actions which can get lost over time. The minimum requirement is deconfliction. The bigger goal is synchronization.[254]

USSOF engagement in Thailand employs campaign logic that respects the sovereignty and sensitivities of the Thai government. This manner of operational art has to skillfully apply US power to expect US influence in return.[255]

The Department of State has the task of safely steering vital US military engagement programs through periods of political upheaval. In the 2006 Thailand coup d'etat example, emerging frictions were managed with few disruptions to US diplomatic relations or military programs. The sustained

---

253   Author interview A19 with Army Special Forces Colonel Paul Ott, September 5, 2012. Colonel Ott commands the Special Warfare Education Group (SWEG) at the United States Army John F. Kennedy Special Warfare Center and School (USAJFKSWCS) at Fort Bragg, NC. Annually, the SWEG language school will graduate between 20 to 30 Thai language educated ARSOF officers per year.
254   Author interview A20 with US Army Special Forces officer, September 06, 2012. This SOF officer explained the role of a SOF in-country representative writ large, not specific to Thailand.
255   Nye Jr., "The Future of Power," 32-50, 225-231.

US-Thai SOF engagements exemplify the calibrated balance of assisting a regional partner while pursing US interests. As the US rebalances its strategic focus on the Asia-Pacific region, the US engagement model in Thailand presents a compelling case study on quietly pursuing US interests and building alliances without initiating provocative maneuvers against potential competitors or threats.

Summary: Policy and Engagement

Operational art in a non-wartime environment aspires to translate policy into programs, actions, and engagements that achieve strategic objectives. The crafting of operational art in a Phase Zero campaign is conducted not *in spite of* policies, but in support of them. USSOF engagements or putative campaigns exist within imperfect, even byzantine, policy arenas. The vignettes of Yemen (*incremental* policy), Leahy Act (*legislative* policy), and Thailand (*political* policy) present the different circumstances where SOF uses and adapts operational art to achieve coherency. These vignettes illustrate the tensions of linking tactics to strategy without the liberties often inherent in a declared theater of war.

In Phase Zero, host nation sovereignty and US policy are paramount. Subsequently, operational art becomes a subordinate discipline. This limits military flexibility and restricts freedom of action. If sluggish bureacratic procedures are also in play then the impediments to operational art become overwhelming. Overcoming these obstacles to preserve the coherency of USSOF engagements is the key challenge. The Yemen, Indonesia, and Thailand vignettes show how such pressures impacted USSOF expressions of operational art. COL (Ret.) Kevin Benson, a former director of the US Army School of

Advanced Military Studies (SAMS), summarizes the need to develop an operational art scheme in this type of Phase Zero environment.

> The art of strategy is giving politically aware military advice to policy makers. Clausewitz cites that war and the use of force is an extension of policy – this is where the operational art comes in ... we need operational artists in Phase Zero who can rationalize and synchronize engagements to inform the current operations folks on what we are pursuing, what policy we are supporting, and why this is important strategically. No matter what we call it, this is about understanding and supporting strategic and policy objectives.[256]

Where military actions are restrained and potentially counterproductive, the contemplation and application of a creative, non-standard operational art construct takes on even greater importance. While this is simple in concept it is difficult in practice.[257] Extracting insights from the USSOF Thailand model is a sound start to reframing operational art constructs suited to these environments.

Section IV. Synchronization Challenge Two: Programs

Whereas policy sets direction, guidance, and parameters, US programs are the implementing mechanisms of policy.

---

256 Author interview A10 with COL (Ret.) Kevin Benson, a former School of Advanced Military Studies (SAMS) Director, August 20, 2012.

257 Christopher J. Lamb, "Statement of Christopher J. Lamb, Distinguished Research Fellow, Center for Strategic Research, Institute for National Strategic Studies, National Defense University on 'The Future of U.S. Special Operations Forces,'" before the Subcommittee on Emerging Threats and Capabilities, House Armed Services Committee, U.S. House of Representatives, July 11, 2012.

Programs refers to the US government funded international programs binned under the broad title of *Security Cooperation*.

Security Cooperation combines "all DoD interactions with foreign defense establishments to build defense relationships that promote specific US security interests, develop allied and friendly military capabilities for self defense and multinational operations, and provide US forces with peacetime and contingency access to a host nation."[258] It consists of seven program types with multiple sub-programs under each category. In all, US Security Cooperation consists of a staggering 97 programs or sub-programs of record (Figure 12).[259] The seven main programs associated with the Department of Defense are security assistance, global train and equip, international armaments cooperation, humanitarian assistance, training and education, combined exercises, and military-to-military contacts.

These programs are the strategic *means* of delivering US security capacity to foreign partners. The construction of a country engagement plan interprets strategy goals, policy intent, and military objectives and conceptualizes an operational approach appropriate to accomplish these ends. Shifting from art to science, implementation requires aligning resources, authorizations, approvals, and programs to execute the operational approach.[260]

---

258 DISAM Greenbook, 1-1. Definition first approved in JP 1-02, 09 June 2004.
259 DISAM Greenbook. The total count of 97 programs is compiled from the DISAM Greenbook.
260 Authorities is a term used to describe statutory and delegated powers. Andru E. Wall describes how authorities are derived in his article, "Demystifying the Title 10-Title 50 Debate: Distinguishing Military Operations, Intelligence Activities & Covert Action," Harvard National Security Journal, Volume 3. "Title 10 of the US Code created the Office of the Secretary of Defense and assigned the Secretary of Defense all "authority, direction and control" over DoD, including all subordinate agencies and commands (10 U.S.C., 113(b)). The President, in his role of Commander in Chief, may delegate through the Secretary of Defense additional responsibilities or "authorities" to US Special Operations Command (USSOCOM). These statutory and delegated responsibilities fall

Security assistance (DoD)	• 15 sub-programs
Global train and equip	• 18 sub-programs • Includes "1206" Building Partner Capacity of Foreign Militaries & "1208" Support of Special Operations to Combat Terrorism
International armaments cooperation	• 8 sub-programs
Humanitarian assistance (HA)	• 9 sub-programs
Training and education	• 9 sub-programs
Combined exercises	• 5 sub-programs • Includes Joint Combined Exchange for Training (JCET)
Military-to-military contacts	• 12 sub-programs • Includes State Partnership Program (SPP)

Figure 12. Department of Defense Security Assistance Programs

Within the strategy aims and policy parameters, military engagement must employ suitable programs to pursue objectives. Returning to Clausewitz, these programs form a type of grammar for Phase Zero. Arranging and harmonizing these programs constitutes the tactics. The skilled arrangement of such tactics with the proper strategic understanding arguably requires competencies that are beyond standard professional

---

under the general rubric of "authorities," 87. Thus, authorities determine whether one has the statutory or delegated authority to conduct the activity in question. Secondly, an "approval" connotes concurrence of said activity from the requisite military or civilian leaders overseeing military activities.

military education and training. Implementing these programs in a sustained and synchronized manner is a challenge that requires tremendous knowledge outside of one's professional domain knowledge. Even when Phase Zero programs are implemented with great skill, tensions arise when reasonably sound program implementation produces negative effects. A former Commander of US Pacific Command, Admiral (Ret.) Timothy Keating, reflected on this lesson from a humanitarian assistance mission.

> I wasn't in the [PACOM Commander] job two weeks and we had just deployed big white ships, big red crosses to Indonesia and engineers, doctors, nurses, and veteraniarians had gone ashore to take care of all manner ailments … as so as I departed the plane I'm prepared to bask in the glory following this visit …the verbal bouquets came but they were somewhat muted and I was surprised by this. Finally, an [Indonesian] Ministry of Foreign Affairs guy pulled me aside and said, 'Listen, we really appreciate the big white ship with a big red cross but next time please find another way to get your doctors, your nurses, your engineers here because our people look at this [ship] as an unmistakable, visible, muscular sign of US presence. And they like what they see but then they look to the leaders of

Indonesia – democratically elected leaders – and they see weakness on our part because we cannot provide [this]."[261]

Admiral Keating humanized the subtleties involved in achieving the right effect when executing humanitarian assistance, and by extension, security cooperation events. Each of the 97 security cooperation programs has a lead agency for approval, execution, and oversight. In the case of security assistance, seven of its twelve programs are administered by DoD through the Defense Security Cooperation Agency (DSCA) "under the general control of the Department of State" as a component of Foreign Assistance.[262] These Department of Defense administered security assistance programs contain complex budgetary procedures, logistics, and other detailed programmatics. As a result, these security assistance programs are often managed by logistic directorates (J4) within the combatant commands.[263] This byzantine system creates unfortunate byproducts: the chain of delivery for engagement can be extensive, bureaucratic, and extended so far across time, space, and organizations that the results are diluted, or worse, manipulated.[264] In this context, simply conceptualizing operational art is difficult enough. Implementation entails yet another level of complexity. The intended effects of Security Cooperation can easily get confused in these environments.

For special operations activities, Phase Zero programming

---

261 Timothy Keating, Center for Strategic and International Studies (CSIS), "US Forward Presence in the Asia-Pacific Region," audio captured from panel discussion, September 24, 2012, accessed on iTunes on January 28, 2012, 1:10:00 to 1: 21:00.
262 DISAM Greenbook, 1-1.
263 US Pacific Command is one example where security assistance is executed under the auspices of the J4.
264 Author interview A04 with an Asia-Pacific experienced Special Forces officer, August 10, 2012.

is focused on a smaller number of these 97 programs. Over 50% of USSOF engagements in Phase Zero occur under the Joint Combined Exchange for Training (JCET) program.[265] JCETs are usually four to six weeks in duration with intimate tactical interactions between USSOF and host nation partners. Known as the workhorse of USSOF, the JCET itself is an "ill-suited mechanism to build partner capacity and capability."[266] A former Theater Special Operations Command operations officer, Colonel Greg Wilson, plainly stated, "JCETs are inadequate tools to build capacity. A new platform is needed."[267] JCETs retain their utility because they are convenient to program and resource for both USSOF and host nation units. Though legally used for such purposes, the JCET is a legacy construct that is crafted to principally benefit the US special operations unit in pursuit of regional familiarization and secondly benefit the partner hosting the exchange.[268] The principle shortcoming is that JCETs are episodic events that are subject to wide variations in host nation hosting units, regions, and desired capabilities. At their best, JCETs are flexible, short duration engagements that increase USSOF regional knowledge, improve host nation specific skills (counterterrorism, peacekeeping, small unit tactics), and strengthen relationships between participating personnel. With a clear notion of operational art and proper resourcing, JCETs can be integrated into a reliable vehicle that successfully links engagements to strategies. At their

---

[265] The figure of 50% is the author's estimate given the frequency of the JCET in light of other SOF venues, across all COCOMs. As described in the Thailand vignette, it is difficult to enumerate exact percentages given that engagement counting by event can portray a false picture of volume, capacity and effect. Numbers aside, the JCET is and remains the primary tool of USSOF to engagement host nation counterparts in Phase Zero environments.

[266] Author interview A21.

[267] Interview with US Army Special Forces officer, Lieutenant Colonel Adrian Donahoe, August 20, 2012.

[268] DISAM Greenbook, 1-20.

worst, JCETs are single events, unrelated to broader strategic aims, providing perishable tactical exchanges with little follow-on actions. Still worse, JCETs can be manipulated by host nation governments or militaries to accomplish ends sparsely related to mutual security goals. With a programmatic cycle of up to two years prior to execution, JCETs can be thoughtfully conceptualized as the former but easily end up as the latter.[269]

Photo 8. JCET programs are the main vehicle for USSOF Phase Zero engagements with friendly nations (US Army photo, authorized use).

Formulating operational art for Phase Zero is possible using blended approaches through uniquely suited programs. The numerous US Phase Zero programs do provide certain strategic advantages. The wide range of programs and authorizations available to accomplish military engagement objectives

---

[269] Author experience in planning, programming and executing JCETs in the US Army Special Forces, 1997-2010.

are vast, well-funded, and span across multiple security disciplines and domains. However cumbersome or ill-suited these programs may be to singularly formulate operational art required for Phase Zero operations, they do offer complementary means and methods to build engagement campaigns within host nations. In a well-conceptualized strategy, the skilled application of operational art involves blending the programs from different authorizations to include Title 22 (DoS), Title 10 (DoD) and Title 50 (Intelligence).[270] Summarizing this challenge, an Asia-Pacific based special operations planner stated, "Yes, this is often a patchwork job with a few big wins and potentially a few dead ends. But often the missed opportunities are often from lack of imagination and creativity as much as the overly complicated programmatics."[271]

Section V. Synchronization Challenge Three: Posture

The joint US military forces' global posture evolves according to strategic aspirations, perceived security requirements, and current and emerging threats of the United States. Military posture is more than just forward bases, logistics, and personnel. It also includes "relationships, activities, facilities, legal arrangements, and global demands."[272] Military force posture, reported to Congress annually, reflects the military-strategic views on where, how, and why US forces are best positioned within the modern strategic environment.[273]

---

270 Title 50, U.S.C. delineates intelligence responsibilities which includes both the CIA and DoD. Despite the usage, "Title 50" is not synonymous with CIA activities.
271 Author interview with US Army Special Operations officer, LTC Michael Kenny, August 06, 2012.
272 Army Posture Statement (2009), available at www.army.mil/aps/09/information_papers/global_force_posture.html. Accessed on January 10, 2012.
273 Posture statements are issued annually by the armed services (including USSOCOM), and the combatant commands to the committees and subcommittees of the Senate and House of Representatives.

In 500 BC, Chinese strategist Sun Tzu discussed the idea of *shih* or "strategic configuration of power."[274] *Shih* combines the concept of positional advantage with the intangibles of energy, momentum, force, influence, authority, and power.[275] Similarly, US military posture in the Phase Zero environment calculates how to strategically configure US power to optimum effect. Giving context to shih, Strategists Henry Mintzberg and James Bryan Quinn declare, "the essense of strategy – whether military, diplomatic, business, sports (or) political ... is to build a posture that is so strong (and potentially flexible) in selective ways that the organization can achieve its goals despite the unforseeable ways external forces may actually interact when the time comes."[276] In non-wartime situations, engagements carry a heavy burden for the achievement of strategic objectives. Thus, force *posture* becomes just as important as force *employment*.

USSOF Phase Zero posture can be problematic. According to USSOCOM, the current system of posture, authorities, and force allocation is ill-suited to today's security environment.[277] USSOCOM is seeking more flexibility to anticipate, deploy, and support appropriate special operations activities up to and including special operations-centric campaigns.[278] In a 2012 testimony to the House Armed Services Committee,

---

274 Sun Tzu, *The Art of War*, translated by Ralph D. Sawyer (Boulder, Colorado: Westview Press, Inc.) 1994, 142-145. Original text of *The Art of War* is attributed to the 500 BC time period. While there is no direct English language translation of "shih," Sawyer cites the elements of shih as "circumstances, energy, latent energy, combined energy, shape, strength, momentum, tactical force, power, authority, influence, power, condition of power, force of circumstances, positional advantage, and purchase," 144.
275 Tzu, *Art of War*, 143.
276 Mintzberg and Quinn, *Readings in the Strategy Process*, 8.
277 Eric Schmitt, "Elite Military Forces are Denied in Bid for Expansion," The New York Times, June 4, 2012, http://www.nytimes.com/2012/06/05/world/special-ops-leader-seeks-new-authority-and-is-denied.html?hp&_r=0, (accessed December 15, 2012).
278 Eric Schmitt, "Elite Military Forces are Denied in Bid for Expansion," The New York Times, June 4, 2012, http://www.nytimes.com/2012/06/05/world/special-ops-leader-seeks-new-authority-and-is-denied.html?hp&_r=0, (accessed December 15, 2012).

Linda Robinson, a senior fellow on the Council of Foreign Relations, discussed the potential changes.

> What is needed is greater agility in the review and approval process [for security assistance], since it can take up to two years in some cases... [another] major proposal tabled by USSOCOM would explicitly give USSOCOM a global area of responsibility, allow it to initiate requests for forces, and via a global employment order allow USSOCOM to shift assets among theaters with the concurrence of the geographic combatant command.[279]

Using Bryan and Mintzberg's concept, USSOCOM is focused more on the *flexibility* aspect and less on raw *strength*. To improve special operations global agility, USSOCOM is proposing new methods to "integrate the steady-state with the contingency."[280] Proposed methods include revisions to the Unified Command Plan and how forces are assigned, apportioned, and allocated to theaters of operation.[281] In order to improve special operations campaigning in support of combatant commanders, a USSOCOM representative stated,

> It is about posture and authorities ... the ability to conduct activities beyond that which is only threat-centric ... something with the qualities of a deployment order

---

279 Robinson, "Testimony on Special Operations Forces," 5-6.
280 Author visit to USSOCOM in December, 2012. Quotes taken from an unclassified but sensitive briefing on expanding the global SOF posture under Chatham House rules.
281 JP 1-02. Assigned is "to place units or personnel in an organization where such placement is relatively permanent," 23. Apportionment is "in the general sense, distribution of forces and capabilities as the starting point for planning," 19. Allocation is "distrobution of limited forces and resources for employment among competing requirements," 14. These terms have extensive meaning and impacts on force posture. Further discussion, beyond this research project, requires a higher level of classification.

and an execution order that better enables the development of partnerships, establishing persistence presence where required, and posturing the greater SOF enterprise to support if needed ... to do so, we need a fresh look at thresholds and permissions.[282]

USSOCOM seeks to change the systemic force posture and employment paradigms to suit small, regionally aligned special operations teams. Their intent is to improve the authorities under which USSOF teams assess environments and, if appropriate, employ capabilities.[283] Linda Robinson stated, "It requires placing SOF teams out in troubled regions for extended periods so they can gain familiarity, knowledge, and relationships and then begin to execute solutions with the resident partners. This runs counter to the common tendency to wait until crises are full blown and action is imperative."[284]

Special operations seeks improved force posture for both special warfare and surgical strike options. This means more agility to move intra- and inter-theater and expanded authorities to engage partners or irregular forces. This request raised tensions within DoD, DoS, and Congress about giving USSOF wider global latitude. In May 2012, USSOCOM's initial request for revised authorities was rejected by the Congress and DoS.[285] In April 2012, Admiral (Ret.) Timothy Keating said "I don't fundamentally understand what needs fixing."[286] In early 2013, some changes were implemented. Principally, USSOCOM assumed combatant command over theater-based

---

[282] Author USSOCOM visit, December 2012.
[283] Author interview A21.
[284] Linda Robinson, testimony to the House Armed Services Committee threats subcommittee meeting, July 11, 2012. Quote above is referenced from Sandra I. Erwin, "Special Operations Seeks Bigger Role in Conflict Prevention," (referenced earlier).
[285] Schmitt, "Elite Military Forces are Denied in Bid for Expansion."
[286] Schmitt, "Elite Military Forces are Denied in Bid for Expansion."

special operations forces, thus giving USSOCOM greater flexibility in force posture and employment options.[287]

The September 11, 2012 attack on the US Consulate in Libya resulting in the death of Ambassador Chris Stevens brought greater visibility to the current posture and authority debate. An initial Department of State investigation was mute on the topic of how USSOF responds to these types of crisis events.[288] Others have called for an expanded investigation that looks beyond on-site facilites and security.[289] Discussing the absence of USSOF in the Benghazi incident, Lieutenant General John Mullholland, Deputy Commander of USSOCOM stated, "Those [regional USSOF] forces worked as advertised. They were in place."[290] Unstated, but implicit in these remarks, is that the current method of posturing and employing USSOF is ill-suited to capitalize on USSOF capabilities. Ultimately, USSOF teams were not alerted or postured in a manner quick enough to respond. [291] "To be relevant in this security environment, you have to be ahead of the crisis," Mulholland stated.[292] If military force posture truly reflects the military-strategic

---

[287] The revised command arrangement deemed USSOCOM as the combatant commander of regional SOF (TSOCs). However, regional SOF will remain under the operational control (OPCON) of the Geographic Combatant Commanders. This shift allows USSOCOM to assume responsbility for the TSOCs as well as provide more responsive resourcing of SOF to Geographic Combatant Commanders. The main concern from combatant commanders was the fear of losing SOF to other theaters, on short notice, that will disrupt their Theater Campaign Plan execution.

[288] Department of State Accountability Review Board (ARB) chaired by Ambassador Thomas R. Pickering, unclassified report invesigating the 11-12-2012 Attack on the US Consulate in Benghazi, http://www.scribd.com/doc/118488083/State-Department-Investigation-of-Benghazi-Attack-of-9-11-2012 (accessed January 28, 2013), 1-39.

[289] Sarah Parnass, "Hilary Clinton Endures Brusque Questioning at Hearings," ABC News online, January 23, 2013, http://abcnews.go.com/Politics/OTUS/hillary-clinton-endures-brusque-questioning-hearings/story?id=18292329 (accessed January 29, 2013).

[290] Erwin, "Special Operations Seeks a Bigger Role," 1.

[291] Accountability Review Board. The unclassified report does not fault USSOF. Reportedly, USSOF regional response teams were eventually positioned at Sigonella Air Base, Sicily. The timeline for alerting and posturing USSOF is not available in open source documents.

[292] Schmitt, "Elite Military Forces are Denied in Bid for Expansion."

GOING BIG BY GETTING SMALL

views on where and how US forces move and act, then current paradigms need evaluation. This current debate shows the tensions of shifting US power projection paradigms to provide greater latitude for USSOF operations.

Photo 9. Skilled partnered force training exchanges employ methods suitable and sustainable for partnered nation forces (US Army photo, authorized use).

Conclusion: Phase Zero Tensions

This chapter charted the thicket of Phase Zero interests and tensions. Phase Zero is a mosaic of intent and action in a non-wartime environment. US policies, programs, and posture sets the theater for the pursuit of US interests. No less important are unique organizational cultures and the peculiarity of each host nation. Within this environment, special operations pursues direct and indirect power projection methods that operate within defense, diplomacy, and development realms. Adding further complexity is the extended timelines of Phase Zero operations, actions, and activities. The extended timelines degrade unity of purpose and challenge continuity schemes. In Phase Zero environments, special operations operational art aspires to devise the right access, understanding, and engagement venues. Frictions are inherent. Key to overcoming these frictions and tensions are operational concepts that view all forms of art – diplomatic, military, developmental, and social – as necessary for the attainment of US strategic interests. The case study of USSOF actions in The Republic of Colombia presents one such example of the successful fusion of US governmental power in support of US interests vis-à-vis our Colombian partners.

4
Case Study

Colombia's turnaround from a 1990s near-failed state to one of the most stable governments in South America is foremost a Colombian story.[293] A sub-narrative of this story is the United States' support to the Colombian political and military reforms over the past forty years. This case study focuses on the role of US special operations from 1998 to 2008 within the larger US government support to Colombia (Figure 13). A close examination of the July 2, 2008 Colombia special operations hostage rescue mission of Colombian and American citizens aids in the analysis of the decade of USSOF persistent engagment. Importantly, this hostage rescue mission is not a culiminating point but rather is an exclamation point within the continuing story of the US-Colombia partnership. The case study themes are on the operational level maturation, development, and architecture of the USSOF effort in Colombia. Examining these themes increases understanding of the components of operational art within a Special Forces Phase Zero campaign construct.

---

[293] Thomas Marks, "Colombia: Learning Institutions Enable Integrated Responses," National Defense University Center for Complex Operations, Prism 1, No. 4, September 2010, 127-146.

CASE STUDY

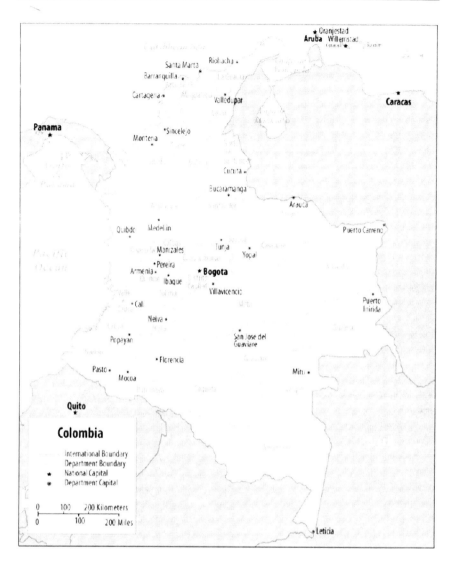

Figure 13. Map of Colombia, CRS, November 2012

To reveal and understand the elements of operational design employed by Special Operations Command South (SOCSOUTH), this case study is organized into four sections.

The first section provides a brief background on Colombian politics, threats, and policies. Second, US policy and combatant command strategy toward Colombia is reviewed. Third, USSOF engagement in Colombia is outlined. Finally, in chapter five, the USSOF campaign in Colombia is analyzed using the thirteen elements of operational design from Joint Publication 5-0, *Joint Operation Planning*.[294]

Background: Colombia

Relations between the US and Colombia took on a new dimension with a 1959 US State Department and Central Intelligence Agency assessment team investigating the Colombian lawlessness known as *La Violencia* (1948-1966).[295] Beset by internal civil strife, insurgency, and murderous violence, Colombia accepted US offers of assistance. In 1962, Brigadier General William Yarbrough, US Army Special Forces, led a counterinsurgency advisory team at the invitation of the Colombian president.[296] From the US assessments and advice came Plan Lazo (Lasso), a country-wide strategy to restore civil order.[297] Decades of strife and insur-

---

294 This case study uses the term "SOF campaign" to describe the extended USSOF engagement in Colombia. By definition, the totality of USSOF efforts commanded by a joint SOF headquarters (SOCSOUTH) clearly meets the joint definition of a campaign, "a series of related major operations aimed at achieving strategic and operational objectives within a given time and space" (JP 5-0, 37). The term "sub-campaign" could also be applied, though the term "subordinate campaign plan" has doctrinal connotations to service or sub-unified plans that support a DoD global campaign plan (JP 5-0, II-27). Critically, the main campaign in play is the Colombian campaign. However, to distinguish the uniquely architected USSOF effort in support of the Colombian campaign, and subordinate to a combatant command theater strategy, the term "SOCSOUTH campaign" is used.
295 Troy J. Sacquety, "Colombia's Troubled Past," *Veritas: Journal of Army Special Operations History* 2, no. 4 (November 2007): 8-14.
296 Sacquety, 9.
297 Dennis M. Rempe, "The Past as Prologue: A History of the US Counter-Insurgency Policy in Colombia 1958-1966," March 2002, Strategic Studies Institute Monograph, www.strategicstudiesinstitute.army.mil/pdffiles/PUB17.pdf (accessed January 17,

gent violence continued in Colombia, with a sustained civil war against left-wing opposition groups led by the *Fuerzas Armadas Revolucionares de Colombia* (FARC), the *Ejercito de Liberacion Natcional* (ELN) and the right-wing *Autodefensas Unidas de Colombia* (AUC).[298] Compounding Colombian political instability was the increase in drug production and profiteering from a growing international demand for narcotics.[299] With Colombian security hemorrhaging and an increasing US domestic drug problem, US engagement in Colombia shifted. By the 1990's, US engagement in Colombia was centered on counternarcotics.[300]

In 1999, Colombian President Andres Pastrana Arango presented a comprehensive civil, military, and development plan called Plan Colombia. The Department of State led the US efforts, with the Narcotics Affairs Section (NAS) of the US embassy in Bogota delegated a large role. Though the plan contained a comprehensive mandate on all civil sectors, Pastrana's Plan Colombia was criticized as overly focused on military and police infrastructure and operations.[301] With its $7.5 billion pricetag, the success of Plan Colombia (1999-2005) is still a topic of debate.[302] Despite its shortcomings, the plan was an impetus for positive change. A noted authority on Colombian counterinsurgency, Dr. Thomas Marks, stated

---

2013).
298 Jane's online, "Fuerzas Armadas Revolucionares de Colombia (FARC), Ejercito de Liberacion Nacional, and Autodenfas Unidas de Colombia (AUC)," January 17, 2013, www.janes.com.
299 Geoff Simons, *Colombia: A Brutal History* (London: Saqi Books, 2004).
300 Robert W. Jones, Jr., "Plan Colombia and Plan Patriota: The Evolution of Colombia's National Strategy," *Veritas* 2, no. 4, 60-64.
301 Andres Pastrana, *Plan Colombia: Plan for Peace, Prosperity, and Strengthening the State*, Government of Colombia, Office of the Presidency, October 1999 (English version), http://www.presidencia.gov.co/webpress/plancolo/plancin2.htm.
302 US GAO, Plan Colombia: Drug Reduction Goals Were Not Fully Met, but Security Has Improved; US Agencies Need More Detailed Plans for Reducing Assisstance, October 2008, GAO-09-07.

that Plan Colombia reclaimed "the strategic initiative" for the Colombian government in its fight to contain both the insurgency and the drug trade.[303] Pastrana's successor, Alvaro Uribe Velez (2002-2010) seized on this momentum. President Uribe succeeded Plan Colombia with Plan Patriota, a muscularized version of Plan Colombia that capitalized on the expanded Colombian security capability.[304] Initiated in 2004, Plan Patriota was the military component of a dual security strategy designed to secure rural populations and drive the FARC from their safehavens (*despeje*).[305] Uribe also ushered in a change in the strategic use of Colombian security forces. Uribe explained "The original division was for the military to protect our borders and the police to secure our cities. We couldn't stick to this. We had to involve the military in a fight against narcotrafficking, and we had to involve the police in our fight against terrorist groups such as guerillas and self-defense groups."[306] Under the Uribe administration, Colombia achieved tremendous progress in building effective military and police forces, quelling its decades-old insurgencies, and restoring civil order and confidence.[307] The results reduced the threat of Colombia falling into a narco-state dominated by the powerful Cali and Medellin cartels.[308] Punctuating this success was a dynamic, sophisticated hostage rescue operation on July 2, 2008, conducted by Colombian special operations units. The daring mission rescued three US and twelve

---

[303] Thomas A. Marks, "A Model Counterinsurgency: Uribe's Colombia (2002-2006) vs. FARC, Military Review, March-April 2007 *Military Review*, http://usacac.army.mil/cac2/call/docs/11-15/ch_6.asp (accessed January 17, 2013).

[304] Plan Patriota was one component of Uribe's overall reform strategy entitled, *Democratic Security Policy and Plan* (DSDP).

[305] Jones Jr., "Plan Colombia and Plan Patriota," 63-64.

[306] Alvaro Uribe Velez, "An Interview with Alvaro Uribe Velez," *Prism* 3, no. 3, June 2012, (Washington DC: National Defense University Press), 141.

[307] June S. Beittel, "Colombia: Background, US Relations, and Congressional Interest," Congressional Research Service, November 28, 2012, 1-26.

[308] Beittel, "Colombia," 18-26.

Colombian citizens. The Americans were held in FARC jungle camps for over five years.[309] The progress achieved in Colombia from 1998 to 2008 is startling. Colombia still faces stability challenges. However, modern day Colombia is no longer the near-failed state it was in 1998.[310] In 2009, US Ambassador to Colombian William Brownfield stated that Colombia "has been the most successful nation building exercise that the USA has associated itself with in the past 25 to 30 years."[311] Shaping this process was US support led by a significant USSOF effort.

US Southern Command and US Policy

Illegal drugs, principally cocaine and heroin, drive US interests in Colombia. The scourge of South and Central American narcotics in US cities and streets remains the locus of US policy towards Colombia.[312] Aside from the US appetite for drugs, the South and Central American drug trade has evolved into transnational organized crime (TOC) networks with sophisticated methods that threaten the rule of law and modern state security capabilities.[313] To support US security objectives, the Miami, Florida based US Southern Command devised four regional strategies. The four regions are the Andean

---

309 Juan Forero, "In Colombian Jungle Ruse, US Played a Quiet Role; Ambassador Spotlights Years of Aid, Training," Washington Post, July 9, 2008, http://www.washingtonpost.com/wp-dyn/content/story/2008/07/08/ST2008070803342.html?pos=, (accessed January 17, 2013).
William Brownfield was the US Ambassador to Colombian from September 12, 2007 to August 5, 2010.
310 Marks, "A Model Counterinsurgency," 127-140.
311 CBS Evening News, "Colombia to Aid US in Taliban Fight," July 27, 2009, http://www.cbsnews.com/8301-18563_162-5192173.html (accessed January 28, 2013).
312 Beittel, CRS report, 1-4.
313 Michael Kenney, *From Pablo to Osama: Trafficking and Terrorist Networks, Government Bureaucracies, and Competitive Adaptation*, (University Park, PA: The Pennsylvania State University Press, 2007).

Ridge (Colombia, Venezuela, Bolivia), Central America, the Caribbean, and the Southern Cone (Chile, Argentina, Brazil, Uruguay, Paraguay).[314] Southern Command's challenge was the lack of US armed force capacity to sustain and grow security initiatives. With a relatively small percentage of permanently assigned US military force structure and budget, Southern Command was, and is, an "economy of force theater ... This historically forces tough choices about where to put our strength and where to accept risk."[315]

From 2000 to 2010, Colombia was the third largest recipient of US foreign assistance, trailing only Israel and Egypt.[316] During this decade, the US committed $7 billion in State and Defense Department programs.[317] The preponderance of this aid was counternarcotics funding through the Department of State and the Bureau of International Narcotics and Law Enforcement Affairs (INL).[318]

In the aftermath of 9-11, US counterdrug efforts in Colombia were energized by an expanded charter within the existing (mostly counterdrug) programs. In testimony before the House Armed Services Committee in March 2003, SOUTHCOM Commander General James T. Hill stated,

> ... Congress gave us an expanded authority to use

---

314 General Bantz J. Craddock, "Posture Statement of General Bantz J. Craddock, Commander, US Southern Command, before the 109th Congress House Armed Services Committee," March 16, 2006.
315 Author visit to US Southern Command, December 2012. Author discussion with US Southern Command leadership under Chatham House rules. SOUTHCOM was, and still is, an economy of force theater. SOUTHCOM is the smallest of the six geographic combatant commands in both headquarters personnel and forces allocated.
316 Vaughn's Summaries, "US Foreign Aid Summary 2001-2010," http://www.vaughns-1-pagers.com/politics/us-foreign-aid.htm#top-recipient-countries, (accessed January 28, 2013). US aid to Colombia reduced after 2008. Currently, Colombia out of the top ten of foreign aid recipients.
317 Beittel, "Colombia," 36.
318 Beittel, "Colombia," 27-46.

counterdrug funds for counter-terrorism missions in Colombia because it concluded that there is no useful distinction between a narcotrafficker and his terrorist activity, hence the term narcoterrorist ... Operations today are more efficient and effective because our expanded authorities allow the same assets to be used to confront the common enemy found at the nexus between drugs and terror.[319]

This policy shift allowed the combatant commander wider latitude in applying the full resources of SOUTHCOM against threats to Colombian stability and US interests. For special operations engagement in Colombia, the implications were profound.

US Special Operations Forces Engagement

The evolution of special operations engagement in Colombia from 1998 to 2008 demonstrates how persistent engagement can transform from episodic activities to sustained tactical capacity building and long-term institutional reform. This case study provides one model for a Phase Zero campaign that is SOF-centric, and how aspects of operational art were adapted and incorporated into achieving success.

In the 1990s, US Special Forces engagement with the Colombian military was prohibited due to US policy objections over Colombian human rights issues.[320] During this

---

319 General James T. Hill, "Statement before the House Armed Services Committee on the State of Special Operations Forces," March 12, 2003, http://armedservices.house.gov/comdocs/openingstatementandpressreleases/108thcongress/03-03-12hill.html (accessed January 17, 2013).
320 Robert D. Ramsey III, "From El Billar to Operations Fenix and Jaque" Occasional Paper 34, Combat Studies Institute (Fort Leavenworth, KS: Combat Studies Institute Press, 2009), 18-30.

period, US assistance was provided by US agencies such as the Drug Enforcement Agency and the Department of Justice. Nearly all tactical assistance went to Colombian counterdrug units in the Colombian National Police (CNP). Concurrent with the issuance of Plan Colombia in 1999, the US military began a cautious engagement with elements of the Colombian military.[321] In the 1999-2001 period, USSOF engagement focused on the Colombian National Police counterdrug brigade, the *Brigade Contra el Narcotrafica* (BRCNA) and the Colombian Army Tactical Retraining Center (CERTE).[322] US policy prohibited forces from engaging in counterguerilla activities or operations.[323] All engagement remained counterdrug focused. Concurrent with this expanded USSOF engagement, the FARC, ELN, and AUC moved from guerilla tactics to a war of movement against the Colombian government.[324] During the 1998-2002 period, the Colombia military was reorganizing and reforming itself while engaged in a pitched battle for control of Colombia.[325]

In January 2002, under the aegis of Plan Colombia, USSOF began engagements outside of the counterdrug-focused BRCNA elements. Consequently, USSOF assisted Colombian units that were increasingly operating in the FARC-dominated Southern Colombia regions.[326] This shift brought increased responsibility to the US Army Special Forces Advanced Operational Bases (AOB) headquarters responsible for providing advisory support to Colombian units

---

321 Ramsey, "El Billar," 42-50.
322 Author interview A25 with US Army Special Forces Officer (7th Special Forces Group), September 26, 2012.
323 Ramsey III, "El Billar," 48-50, 69-70.
324 Sacquety, "Forty Years of Insurgency: Colombia's Main Opposition Groups," *Veritas* 2 no. 4, 47-51.
325 Ramsey III, "El Billar," 43-70.
 Uribe Velez, "Interview," 139-144.
326 Author interview A25, 26 September 2012.

without engaging US personnel in combat.[327] A US Army Special Forces officer recalled, "you could see a plan, a linkage, a consistency of effort throughout the training iterations ... tactically, there was a very high level of skill."[328] With an expanded US presence within Colombia, USSOUTHCOM placed Planning and Assistance Training Teams (PATT) within tactical and operational Colombian military and police headquarters.[329] The PATTs addressed the symptoms of a larger problem: the US assistance infrastructure was not designed to manage the totality of programs, equipment, and operational support garnered from the generous US assistance package to Plan Colombia.[330] To address this problem, the US Congress authorized an increase to the limit of US forces allowed in Colombia. In turn, USSOF increased institutional engagements with Colombian Special Operations tactical and operational level headquarters.[331] USSOF supported the creation of a Colombian Army Special Operations Command (*Commando de Operaciones Especiales Del Ejercito* or COESE) and the Colombian equivalent of a Joint Special Operations Command (*Commando Conjunto de Operaciones Especiales* or CCOPE).[332] During this period, the USSOF tacti-

---

[327] Author interviews A25, A26 (US Army Special Forces officer, September 24, 2012) and A11 (US Army Special Forces officer, August 21, 2012). All these officers are from the 7th Special Forces Group with multiple deployments to Colombia during the 1998-2008 period.
[328] Author interview A26, September 24, 2012.
[329] Ramsey III, 107.
[330] Ramsey III, 60-70.
Nina M. Serafino, "Colombia: The Uribe Administration and Congressional Concerns," Congressional Research Service report, June 14, 2002.
[331] Ramsey III, 102. Under the Bush Administration in 2002, the force cap was increased from 500 DoD and 300 contractors to 400 each (800 total), 76. The force cap was again increased in 2004 to 800 DoD and 600 contractors with additional slots authorized for US personnel involved in potential personnel recovery (PR) operations for American citizens, 108.
Author interviews A25 and A26.
[332] Ramsey III, 113-116.
Kenneth Finlayson, "Colombian Special Operations Forces, *Veritas* 2, no. 4, 56-59.

cal, operational, and institutional advisory efforts were conducted by roughly 100 advisors from the 7th Special Forces Group (Airborne) and other joint USSOF units.[333]

Photo 10. Many Phase Zero engagements include small unit tactics (US Army photo, authorized use).

US Hostage Crisis and SOCSOUTH Assessment

On February 13, 2003, three US civilian contractors crash-landed their single turboprop engine aircraft in the Colombian jungle. US citizens Tom Howes, Keith Stansell, and Marc Gonsalves were immediately taken hostage by the FARC.[334] The incident initiated US hostage rescue contingency planning within Colombia. The crisis ushered in a new urgency to the US-Colombian partnership and increased tensions with

---

333 Author interviews A11, A16, A25, A26.
334 Marc Gonsalves, Keith Stansell, Tom Howes with Gary Brozek, *Out of Captivity: Surviving 1,967 Days in the Colombian Jungle*, (New York: HarperCollins, 2009).

the possibility that US unilateral hostage rescue actions might be taken.[335] A US 7th Special Forces Group officer recalled: "The Colombians reacted to the US pressure on finding the hostages ... this generated momentum for better trained units and for a hostage rescue capability. All the pieces were in play to do this."[336] The 7th Special Forces Group led a modest but persistent USSOF rotational presence in Colombia. USSOF engagement remained principally focused on Colombian National Police, CERTE, and CCOPE elements.[337]

In March 2006, USSOUTHCOM Commander General Brantz J. Craddock declared to Congress that his "top priority" was the rescue of the US hostages.[338] General's Craddock's successor, Admiral James Stavridis, echoed this comment in October 2006.[339] In September 2005, an execution order was issued for Operation Willing Spirit. The named operation focused SOUTHCOM's resources and authorities on renewed efforts to locate and recover the US hostages.[340] To synchronize the US effort, SOUTHCOM appointed SOCSOUTH, a sub-unified command, as the lead for all DoD hostage rescue and recovery actions.[341]

The SOCSOUTH Commander, then-Brigadier General Charles T. Cleveland, gauged that an operational level construct was required to manage, coordinate, and synchronize the mosaic of engagement and assistance programs in Colombia.[342] This construct could not involve the large,

---

335 Author interviews A25 and A26.
336 Author interview A26.
337 Ramsey III, 114-115.
338 Brantz J. Craddock, "Posture Statement before the 109th Congress House Armed Services Committee," 16 March 2006, http://ciponline.org/colombia/06031crad.pdf (accessed December 15, 2012).
339 Ramsey III, 128.
340 Author interview A16, A25.
341 Author interview A16.
342 Author discussion with then-Major General Cleveland, March 2012, Fort Leavenworth, KS.

high-visibility signature of a JTF-like US headquarters. The SOCSOUTH Commander stated, "Traditional military structures optimized for unilateral action are neither necessary nor welcome by ambassadors and partner nations."[343] Any USSOF increased capability in support of a US hostage rescue effort would require a *pervasive* but not *invasive* application of US power – one of the distinguising features of special operations power application. SOCSOUTH assessed that a coordinated effort required success in five areas: broadened authorities, appropriate command and control arrangements, combined US-Colombian planning and operations, operational reach into the interior jungles of Colombia, and properly postured hostage rescue forces and enablers.[344] SOCSOUTH viewed the Colombian theater-strategic level as central to this effort. The Colombian operational level joint special operations headquarters, CCOPE, was the proper coordinating headquarters. CCOPE's unique posture harnessed the full capacity of Colombian police, military, and civil capacity to coordinate a complex hostage rescue operation. With this assessment, SOCSOUTH devised its operational approach for all USSOF engagement in Colombia.[345]

SOCSOUTH Operational Approach

The SOCSOUTH operational approach contained four integrated ideas:

---

343 Greg Wilson, "SOF Innovations: Operating in the Gap," unclassified briefing (power point), September 2012.
344 Enablers describes supporting activities and units such as logistics, intelligence, aviation and communication.
345 Author interview A16 with COL Greg Wilson (SOCSOUTH Operations Officer, 2005-2007), August 23, 2012.

1. a creative but transparent use of authorities and forward posture;
2. a tactically focused engagement effort with appropriately missioned Colombian units;
3. a SOC-forward headquarters to provide executive level interface; and
4. personal, private collaboration with strategic leadership including Admiral Stavridis, US interagency leaders, the US Ambassador to Colombia, and key Colombian military and political leadership.[346]

These four ideas could not be fully synchronized by authoritative orders. High trust relationships were the fundamental ingredient for integration.[347] At the tactical level, a 7th Special Forces Group officer recalled, "We had amazing relationships with the Colombians, to include psyop [psychological operations] and civil affairs ... the unknown was usually the friction at the political level ... what was the Colombian capability and figuring out where the US fit into that picture."[348] SOCSOUTH gauged that trusted relationships were adequately in place from the lowest tactical units to the Colombian presidential offices. With the theater set, the joint US-supported Colombian agencies exerted increased pressure on the FARC hostage network.[349]

SOCSOUTH's operational approach coalesced in 2006-2007 commensurate with the Colombian government's

---

346 Author interview A16, August 23, 2012.
347 Author interviews A11, A16, A25, A26.
348 Author interview A11, August 21, 2012.
349 Author interview A16. USSOF adopted a "compel" posture with a maturing Colombian military and police capacity to accept greater risks in locating the US (and Colombian) hostages. A joint task in Phase Zero is to help "set the theater" for the potential introduction of major combat forces. In this case, "setting the theater" means having the appropriate relationships and capabilities in place to accomplish more difficult security missions

consolidated gains against the FARC, ELN and AUC.[350] The operational approach generated a more aggressive and higher risk series of operations to locate the hostages.[351] SOCSOUTH benefited from expanded authorities to situate select US advisors with Colombian reconnaissance units and to participate in the sensitive site exploitation of FARC camps.[352] With expanded Colombian operational reach into the deep jungles, the US hostage trail was located. Soon, the FARC method for holding and moving hostages was discovered.[353] With a hostage rescue attempt becoming more probable, SOCSOUTH increased its combined operations training with the Colombians and conducted joint staging rehearsals for hostage rescue attempts. SOCSOUTH developed options for rescue that included US-only, combined US-Colombian, and Colombian-only force packages. These efforts were coordinated by the SOC-forward, an 0-6 level USSOF command. The SOC-forward consisted of a small, joint staff located in Bogota. The SOC-forward acted as a synchronizing headquarters responsible for assisting the Colombian effort and coordinating the proper level of US assistance.[354] The lightly manned SOC-forward managed the technical support tasks, tactical advisory actions, and the daily coordination tasks with US and Colombian officials. The SOC-forward also directed the "reach back" effort of utilizing US-based resources to support forward intelligence and operations in Colombia. The environment remained rife with uncertainty, both politically and tactically. The SOC-forward model reframed how the

---

350 Beittel, "Colombia," 4-26, 46.
351 Author interview A11, A16, A25, A26.
352 Author interview A16.
353 Author interview A16. The interviewee has first person knowledge of Colombian reconnaissance units who discovered, tracked and analyzed the FARC hostage camp network.
354 Author interview, A16.

CASE STUDY

SOCSOUTH commander could provide operational clarity to the combatant commander, Admiral Stavridis. SOCSOUTH's operational approach was as much organizational as it was operational. The four integrating ideas of authorities, tactical reach, forward command and control, and relationships poised SOCSOUTH to appropriately support Colombian efforts. Culturally astute and interagency-savvy leaders at key nodes ensured flexibility.[355] Disciplined improvisation was required to capitalize on fleeting opportunities.

On July 2, 2008, the Colombian special operations and interagency partners successfully executed a stunning hostage rescue operation named *Operation Jaque* (Checkmate). The operation was predicated on months of jungle reconaissance, intelligence, tactical rehearsals, and contingency planning. The rescue operation itself was concieved and executed solely by Colombians.[356] By using simulated communications between FARC leaders, the Colombians tricked a senior FARC commander into loading the hostages onto a helicopter believed to be FARC controlled.[357] Without firing a shot, fifteen hostages were rescued. Three Americans, former Colombian presidential candidate Ingrid Betancourt, and eleven Colombian military and police were repatriated. Audacious by any measure, the Colombians succeeded in a spectacularly daring deception operation.

Within Colombia, *Operation Jaque* made an immediate strategic impact. The rescue resonated in the Colombian population, similar to the daring US raid in Pakistan that killed Osama bin Laden. Beyond the tactical value, the mission publically displayed an exceptional Colombian special operations capability.

---

355 Burton, "ARSOF in Colombia," 26-33.
356 Ramsey III, 135-137
    Author interview A16, August 23, 2012.
    Ferero, "In Colombia Jungle Ruse," July 9, 2008.
357 Author interview A16, August 23, 2012.

Summary

The effects of the US Phase Zero effort in Colombia were on display in *Operation Jaque*. US materiel and professional support combined with Colombian reform programs created an effective precision-operations capability. In turn, USSOF persistent engagement with Colombian special operations was rewarded with a measurable success. *Operation Jaque* was an anamoly as seldom do Phase Zero engagements result in such decisive and highly visible operations. Without overstating its importance, *Operation Jaque* did provide a measurable marker for a sustained engagement strategy. A Special Forces officer summarizes the meaning of this type of achievement. "The highest praise for a FID [foreign internal defense] effort is when the host nation achieves a level of capability, that, when combined with their local knowledge and language, makes them more effective than [the US] could ever hope to be. This is the holy grail of Special Forces work."[358]

The SOCSOUTH rescue operational approach was designed within a broader SOUTHCOM strategy. The overarching SOUTHCOM Phase Zero strategy provided the strategic focus, resources, and authorities. Within the combatant command framework, SOCSOUTH arranged and executed a sophisticated campaign to recover the US hostages. The SOCSOUTH campaign in Colombian has distinctions and nuances important to developing an operational art concept suited for special operations. Chapter five analyzes the distinctions found in the SOCSOUTH operational approach.

---

358 Author interview with US Army Special Forces Major Russ Ames, September 6, 2012.

5
Analysis Of Special Operations Phase Zero Operational Art

The thirteen elements of operational design are used to examine the SOCSOUTH collective actions that constitute the operational approach. Joint doctrine defines the elements of operational design as "conceptual tools to help commanders and their staffs think through the challenges of understanding the operational environment, defining the problem, and developing the approach."[359] By studying the SOCSOUTH operational approach, observations can be made about special operations-peculiar campaign methods. The outcome of this analysis reveals the special operations unique expressions of the elements of operational design. Very few of these expressions are one-for-one replacements of the existing elements of design. Primarily, they are revisions or modifications derived from the standing joint doctrine elements. However small the modifications might be, they are critical distinctions that convey how special operations campaigns are formed. The utility of this analysis is to accurately capture the logic and appli-

---
359 JP5-0, III-18.

cation of special operations expressions of operational art in Phase Zero. These SOF expressions of operational design present a template that is better suited to visualize and plan special operations activities or campaigns.

Nine of the thirteen elements of operational art show unique special operations expressions. This analysis focuses on these nine elements. The remaining four elements show few unique distinctions when applied within a special operations campaign. These four elements (*objectives, effect, line of operation / line of effort*, and *forces and functions*) are not examined in detail. The nine elements are examined with a three step method: doctrinal review, SOF modifications, and justification. First, the doctrinal definition is reviewed. Second, a special operation expression is identified. These expressions are specified as either replacements for or modifications to the existing elements of operational design. The reason for each modification is explained through the SOCSOUTH case study. Figure 14 summarizes the modifications. Completing this analysis, broader special operations Phase Zero conclusions are drawn.

Joint Doctrine Elements	SO Phase Zero Expressions
1. Termination	• Transition
2. Military End-State	• Position of Continuing Advantage
3. Objectives	• Objectives
4. Effects	• Effects
5. Center of Gravity	• Right Partner, Place, Time (R3)
6. Decisive Points	• Decisive Relationships
7. Lines of Op/Effort	• Lines of Op/Effort
8. Direct & Indirect Approach	• Special Warfare & Surgical Strike
9. Anticipation	• Assessments & Programs
10. Operational Reach	• Access & Location
11. Culmination	• Saturation
12. Arranging Operations	• Arranging Chain
13. Forces & Functions	• Forces & Functions

Figure 14. Special Operations Expressions of Joint Elements of Operational Design

ANALYSIS OF SPECIAL OPERATIONS PHASE ZERO OPERATIONAL ART

Termination

Joint doctrine deliberately lists termination as the first element considered in design. Termination criteria are "the specified standards approved by the President and/or the Secretary of Defense that must be met before a joint operation can be concluded."[360] Effective planning requires an understanding of when to end military operations. Termination criteria are then used to "enable the development of the military end-state and objectives."[361] Termination criteria aim to end military operations on terms favorable to the US.

In USSOF Phase Zero, *transition* is recommended as a replacement for termination. Phase Zero (shaping) is conceptually antithetical to termination. Shaping is a continuous process that aims to prevent the type of decisive military actions that require termination planning. Shaping actions have no terminus; they ebb and flow according to priority and resources. Outside of major contingency plans, termination has little application in Phase Zero.

In the place of termination, the SOCSOUTH Colombia campaign conducted transitions. The 1998-2008 USSOF engagements conducted four types of transitions. First, USSOF conducted tactical transitions among Colombian security units. Engagement started in the late 1990s with *BRCNA* and *CERTE* before transitioning to Colombian infrastructure protection battalions.[362] USSOF transitions continued with the development of a multiplicity of Colombian SOF units including *COESE* and *CCOPE*. SOCSOUTH sought to maximize the strategic utility of USSOF by transitioning from established

---
360 JP 1-02, 296.
361 JP 5-0, III-19.
362 Ramsey, "El Billar," 55-76, 100-125.

units to more specialized capabilities.

Second, USSOF transitioned when funding program authorizations allowed wider latitude.[363] The declared nexus of counterdrugs and terrorism dubbed "narcoterrorism" opened up new possibilities for the use of existing US funding.[364] The subsequent minor shifts in counterdrug funding use created immense operational effects.[365] For example, the fielding and use of rotary-wing aviation for air mobility operations drastically improved the operational reach of Colombian SOF.[366] This enabled deliberate shifts to occur when USSOF tactical engagements reached natural transition points.[367]

Third, USSOF conducted a mission focus transition. The February 13, 2003 hostage crisis gave urgency to the USSOF Phase Zero campaign. The transition toward hostage rescue and precision counterterrorism elevated the training level with a renewed emphasis on complex, joint operations. This shift energized both the Colombians and their USSOF advisors.[368] Due to sustained USSOF engagement, the Colombians were ready for this transition.

Fourth, USSOF transitioned to developing Colombian institutional capability. To ensure long-term effects, advisory efforts shifted from tactical units to operational and institutional level engagements. This final transition set the conditions for a reduction in aid to Colombia. Once the third largest recipient of US aid, by 2010 Colombia fell out of the

---

363 Hill, "Statement before the House Armed Services Committee."
364 Beittel, 36
365 Hill, "Statement before the House Armed Services Committee."
366 Author interviews A08, A11, A16, A25, A26. This is an example of security assistance funding shifts that directly impacted USSOF advisory efforts. Counternarcotics funding originated in sections 1004, 1033, and 124 of the National Defense Authorization Act (NDAA).
    Ramsey, "El Billar," 66, 72. The US aviation support included UH-I, UH-II, UH-60 and C-27 aircraft.
367 Author interviews A08, A11, A16, A25, A26.
368 Author interviews A08, A11, A16, A25, A26.

top ten recipients.[369] Transition criteria, developed in place of termination criteria, propelled the SOCSOUTH operational approach.

Transitions are a key element of USSOF operational design. Whereas termination has little use in Phase Zero, proper transitions ensure tactical actions (and programs) remain connected to the accomplishment of strategic objectives. Transitions occur at the tactical, operational, and strategic-institutional levels. Successful transitions prevent wasted US resources where few benefits are derived from stale or irrelevant engagements. Transitions require keen anticipation, far-sighted programming shifts, and joint ownership with partnered units. Transitions are also easily understood in interagency fora. The plain language of transitions reduces confusion and aids shared understanding. In Phase Zero, well timed and executed transitions ensure measurable forward progress with the host nation.

Military End-State

Military end-state is "the set of required conditions that defines achievement of all military objectives. It normally represents a point in time and/or circumstances beyond which the President does not require the military instrument of national power as the primary means to achieve remaining national objectives."[370] Military end-states allow conceptual linkages to be made between final desired conditions and the approach required to accomplish those conditions.

---

369 Marian Leonardo Lawson, Susan B. Epstein, Tamara J. Resler, "State, Foreign Operations, and Related Programs: FY 2011 Budget and Appropriations," April 22, 2011, Congressional Research Service, http://www.fas.org/sgp/crs/row/R41228.pdf (accessed on January 31, 2013).
370 JP 5-0, III-19.

The recommended replacment for military end-state is ***position of continuing advantage***.[371] For SOCSOUTH, military end-states were useful at the tactical level (e.g., develop military hostage rescue capabilities) but had less utility at the theater-strategic level (e.g., build partnerships).[372] In Phase Zero, finite military end-states collide with the logic of sustained, progressive engagements. A former US Ambassador commented, "I loathe the concept of end-state. It's an OK idea but not for the political dynamic."[373] Still today, SOUTHCOM strategic posture confirms, "Building partnerships is the cornerstone of our strategic approach."[374] The strategic logic of constantly improving a partnership is not well served by language that articulates finite military end-states. This incongruity increases the tension between the military and political arms. Military and political logic are more easily aligned when language plainly acknowledges that Phase Zero is the crafting of a continuous engagement strategy. When the logic is synchronous, political-military collaboration encounters less resistance. Strategist Carl Dolman made a distinction relevant to the language and logic of Phase Zero operational design.

> Victory is a concept [that] belongs wholly in the realm of tactics ... The closer one gets to the battlefield, the more meaningful – and obvious – the measure of victory becomes. Accordingly, as the conceptual scope widens from battle to campaign, from campaign to

---

371 Dolman, *Pure Strategy*, 5. Dolman's concept is discussed in Chapter Two.
372 The SOCSOUTH planning and execution products, to include the military end-state, remain classified.
373 A former US Ambassador addressed the students and faculty of the Command and General Staff College School of Advanced Military Studies (SAMS), under Chatham House rules, August, 2012. The author was present.
374 General Douglas M. Fraser, "Posture Statement of General Douglas M. Fraser, United States Air Force, Commander, United States Southern Command, before the 112th Congress House Armed Services Committee, March 6, 2012, 14.

war, and from war to policy, the more troublesome it is even to determine a beginning, much less an end, to events... the outcomes of battles, campaigns and wars are but moments in the unfolding landscape of politics and history ... this larger focus is appropriate for the strategist, who seeks instead of *culmination* a favorable *continuation* of events.[375]

SOCSOUTH's operational art linked tactical actions to strategic objectives by continuous engagement and improvisational actions gauged to continue Colombian forward progress. This approach is entrepeneurial in spirit and distributed in execution. It capitalizes on the organizational design of placing mature, highly trained, regionally aware, and culturally conversant USSOF teams with the appropriate Colombian partners. SOCSOUTH's ten years of engagement was moved forward more by a well-understood commander's intent than by a clear military end-state.[376] In the place of a military end-state, the SOCSOUTH operational approach sought a position of continuing advantage.[377] This distinction cognitively recognizes that Phase Zero is a continuum where success is measured over time rather than by a decisive end-state. Working toward a position of continuing advantage lacks the clean finality of military end-state. Yet it better captures how Phase Zero operational art is visualized and accomplished.

---

375 Dolman, *Pure Strategy*, 5.
376 Author interview, A16. The commander's intent and end-state were not available in unclassified forms.
377 Dolman, *Pure Strategy*, 4-6

Center of Gravity

A center of gravity (COG) is "a source of power that provides moral or physical strength, freedom of action, or will to act ... COGs exist in an adversarial context in involving a clash of moral wills and/or physical strengths."[378] Of joint doctrine's thirteen elements of operational design, center of gravity is alloted the greatest amount of space (four pages) for explanation.[379]

In joint planning circles, center of gravity debates are a central, if not *the* central, feature of defining the problem and understanding the environment.[380] There is good reason for this. Joint doctrine warns, "a faulty conclusion resulting from a poor or hasty analysis can have very serious consequences, such as the inability to achieve strategic and operational objectives at an acceptable cost."[381] The rich discourse required to conduct a center of gravity analysis is at the heart of effective design. And there is the purely cognitive benefit of intellectually grappling with centers of gravity.

However, using the center of gravity as the organizing principle around which Phase Zero operational approaches are constructed is a questionable method.[382] Even greater doubt exists when the nature of the threat facing the US is a networked, non-state actor. Threats such as Al Qaeda or Latin America drug cartels often lack the characteristics of a Clausewitzean state: a government and a fielded miltary. In

---

378 JP 5-0, III-22.
379 JP 5-0, III-22 to III-25.
380 Author experience from two joint assignments and experience in four geographic combatant commands, 1997-2010.
381 JP 5-0, III-23.
382 Celestino Perez Jr., ed., "Addressing the Fog of CoG: Perspectives on the Center of Gravity in US Military Doctrine," by select faculty of the US Army Command and General Staff College, (Fort Leavenworth, KS: Combat Studies Instiute Press, 2012).

these instances, the center of gravity analysis drives a subsequent planning logic that may not suit the problem. A full debate on the merits of the center of gravity is beyond the scope of this analysis. Obtaining definitive data on US and Colombian perspectives on centers of gravity is also challenging. Center of gravity analyses are often more process than product, with few documents to cite as verifiably authoritative.

The evidence suggests that SOCSOUTH viewed the center of gravity model as useful for discourse but did not arrange the operational approach from a doctrinal center of gravity analysis.[383] In the place of center of gravity, SOCSOUTH engineered their operational approach using the *right partner, right place, right time (R3)* method. R3 was not a one-for-one replacement for center of gravity. But its use drove a distinctive planning process that was not engineered from the singular pursuit of attacking or protecting a center of gravity. A special operations strategist summarized the concern with generating approaches from center of gravity analyses. "Center of gravity and schwerpunkt don't apply well to the systemics of a distributed threat. This is much more about flows and networks."[384]

The center of gravity model was ill suited for SOCSOUTH for two reasons. First, the nature of the threat was a blend of insurgency and transnational organized crime that lacked a single center of gravity. When applied, the doctrinal analyses yielded both friendly and enemy centers of gravity at the tactical, operational, and strategic level. This produced six centers of gravity. If the US had six centers of gravity, the

---

[383] Author interviews A11, A16, A18, A25, A26.
[384] Author discussion and brief with USSOCOM representatives at Tampa, FL conducted under Chatham House rules, December 2012. *Schwerpunkt* is a Clausewitzean term that is often translated as "decisive point." It is sometimes translated from German as "heavy point" or "focal point."

Colombians – who viewed the problem differently - had six more. Dr. Thomas Marks explained that the US and Colombia had differing viewpoints about the nature of the threat.

> The [Colombian] assessment stood in stark contrast the to the US strategic view during the Clinton administration (1992-2000), which it sought to impose on the Colombians. In Clausewitzean terms, the United States saw the drug trade as the "center of gravity" ... The Colombian counterassessment argued that this confused an operational center of gravity with the strategic center of gravity – legitimacy, or the support of the people."[385]

The SOCSOUTH approach was to align threat perspectives with the Colombian viewpoint. COL Greg Wilson, the SOC-Forward Commander in 2008, stated that SOCSOUTH adopted the "Foreign internal defense perspective of shaping your support in line with the Colombians. If you are crafting a campaign which is also a host nation campaign support effort, then you need a common analytical picture."[386] Wilson stated that the SOCSOUTH approach was less threat-centric and was instead focused on the placement of specialized US advisor capabilities at the right locations. Instead of a center of gravity analysis driving the operational approach, SOCSOUTH employed the R3 logic of foreign internal defense: right partner, right place, right time. From this placement, USSOF could understand, shape, and support Colombian perspectives and approaches. The US could also rapidly share information and leverage US support to the Colombians. In the short term, this

---
385 Marks, "Colombia: Learning Institutions Enable Integrated Response," 130.
386 Author interview with COL Greg Wilson, August 2012.

method reduced US control. In the long term, this method provided USSOF deep access and helped create shared understanding with the Colombians. With the right advisors distributed amongst the Colombians in no less than twelve locations, USSOF presence deepened the trust and applied US capacity without an overwhelming presence.[387] This approach yielded greater influence in the long term.

The R3 approach also characterizes the logic of USSOF support in Yemen and Thailand. The tactical expression of R3 – distributing select individuals or small teams– promotes solutions that emanate from host nation institutions and leadership. This model seeks to avoid inadvertently blemishing host nation actions by a visibly robust US presence.[388] R3 accepts risk in direct control but it gains ground in establishing enduring relationships that can better withstand crisis or conflict. R3 is not a replacement for a center of gravity analysis. But it can be a derivative of a center of gravity analysis that drives a different logic for applying US power. R3 seeks to be nimble but assertive when required. By adopting the R3 model, the gap shrinks between the US and the host nation operational approaches.[389] This will not automatically decrease tensions. But it can reduce the misunderstandings that generate unnecessary tensions.

Still today, the US-Colombia partnership is viewed as

---

[387] Author interview A08. The US personnel distributed amongst Colombian bases and outposts varied. The number twelve represents small advisory efforts. The US MILGROUP has a persistence presence in several areas in Colombia to manage security assistance programs. USSOF also has a steady-state presence at locations like the LANCERO school that were initiated in the 1950s.

[388] Author interviews A03 and A08. Both of these Special Forces officers worked FID missions in close coordination with the host nation and US Embassy.

[389] An example of this is the US adoption of counterinsurgency as our main strategy (and tactic) in Iraq and Afghanistan. From an R3 perspective, our mission would be viewed as foreign internal defense *in support of* Iraq or Afghan counterinsurgency. This distinction, while seemingly small, drives distinctly different behaviors and operations.

the cornerstone of the US strategy.[390] If so, then operational design logically starts with how we interface with the Colombians. Whereas maneuver operational art contemplates the emplacment of battalions, artillery, and airfields, USSOF operational art gives the same importance to the selection and placement of single advisors. Its form is different but its aim is the same: posture correctly to shape tactics toward the accomplishment of strategic objectives. Bridging these two different ideas of operational art with improved concepts and language is an important step in successfully integrating USSOF actions with other forms of military power. The logic of R3 is one such bridging method. The adoption of a deliberate method of right partner, right place, right time is a central component of crafting a Phase Zero special operations operational art.

Decisive Points

A decisive point is "a geographic place, specific key event, critical factor, or function that, when acted upon, allows a commander to gain a marked advantage over an adversary or contributes materially to achieving success (e.g., creating a desired effect, achieving an objective)."[391] Joint doctrine ties decisive points to centers of gravity. "Although decisive points are usually not COGs, they are the keys to attacking or protecting them."[392]

In crafting an operational art, the SOCSOUTH emphasis was less on decisive points and more on *decisive relationships*. Decisive points were used not exclusively to attack a

---
390 Fraser, "Posture Statement," March 6, 2012.
391 JP 5-0, III-26.
392 JP 5-0, III-26.

center of gravity. They were used to identify events or areas critical to accomplishing major objectives.[393] Decisive points were employed as a part of the standard tactical and operational-level lexicon and logic. A Special Forces officer describes this quality.

> Most military forces are looking at organic resources and how to apply them. SOF looks at what is in place, local indigenous populations, potential partners, and relationships. Relationships are the beginning of understanding how to openly approach the problem set ... this usually brings indirect solutions or other alternative methods that could not be conceptualized nor could they be used without the right relationships."[394]

The SOCSOUTH relationships from 1998-2008 grew from initial USSOF JCETs into long-term, joint capacity building relationships. Professional relationships grew outward from joint training exchanges conducted with specific Colombian units. From these relationships came the improvisational aspect of USSOF engagement. When relationships are immature or temporary, there are only vague notions of how to link tactics to strategy. When relationships are deepened, the pathways to strategic objectives begin to appear with greater fidelity. A SOF officer stated, "We should have access and relationships where host nation operational art is determined. We need to ascertain what relationships do we want to build, where, and to what purpose. Getting the right person to do this is akin to emplacing a strategic platform."[395] In Colombia,

---
393 Author interview A02, A16.
394 Author interview with US Army Special Forces Lieutenant Colonel Josh Walker, July 18, 2012.
395 Author interview A13, US Army Special Forces officer, August 10, 2012. This particular

strong relationships were established at multiple levels. SOC-Forward Commander COL Greg Wilson stated,

> The hostage crisis galvanized the US country team. Ambassador Brownfield was a tremendous supporter. He was just fine with having three 0-6's in country. Admiral Stavridis gave SOCSOUTH the full support of the COCOM. The Colombians were very willing partners with a long-term commitment. The glue to all this were the distributed, [Spanish-language] fluent US teams. They bridged the gap in the human dynamic when operations got more complex and nuanced.[396]

The SOCSOUTH notion of decisive relationships is a seemingly self-evident idea. However, the theater architecture of plans, policy, and resourcing conforms to a different logic. A SOF officer explained.

> The combatant command is not averse but there is a fundamental prejudice in the system against what SOF does. Military wisdom says that if we have one person in the embassy, then twenty will make it even better. Force multiplication through relationships is not a well understood principle within operational art … Resources are usually allocated and committed according to CONPLANs, when the more valuable asset is a relationship in a key place. It's hard to justify resources for this.[397]

---

interview was not specific to SOCSOUTH or USSOF operations in SOUTHCOM.
396 Author interview. The three O-6's (Colonel or equivalent) were the Defense Attache, the MILGROUP Commander, and the SOC-Forward Commander.
397 Author interview US Special Forces officer, LTC Michael Kenny.

ANALYSIS OF SPECIAL OPERATIONS PHASE ZERO OPERATIONAL ART

Relationships have an ephemeral quality that is hard to qualify within doctrinal concepts or pure physics-based military problems. Yet recognizing relationships as *decisive* expresses their value in familiar military language. It also promotes the pursuit of relations vice just transactions; a vital quality in establishing trust with US and host nation colleagues. Decisive relationships are an expression of operational art that emphasizes posture more than prescriptive action. In Phase Zero, bridging the gap between tactics and strategy is often predicated on relationships. From those relationships come greater possibilites in tactical actions, operational creativity, and strategic effect.

Direct and Indirect Approach

Joint doctrine defines approach as "the manner in which a commander contends with a COG. A direct approach attacks an enemy's COG or principal strength by applying combat power directly against it ... An indirect approach attacks the enemy's COG by applying combat power against a series of decisive points that lead to the defeat of the COG while avoiding enemy strength."[398] This definition distinguishes direct and indirect according to their relationship with the center of gravity.

For USSOF, direct and indirect approach are more clearly expressed by the ***special warfare*** and ***surgical strike*** concepts. USSOF in general, and SOCSOUTH specifically, define direct and indirect approaches using a different context. In a step away from joint doctrine, direct and indirect approaches are not defined *by their relation* to the center of gravity. USSOF define direct and indirect by the manner in which USSOF

---
398 JP 5-0, III-32.

applies power. USSOCOM Commander Admiral McRaven distinguished the approaches in his 2012 Senate Armed Services Committee testimony.

> The direct approach is characterized by technologically-enabled small-unit precision lethality, focused intelligence, and interagency cooperation integrated on a digitally-networked battlefield... However, the direct approach ... only buys time and space for the indirect approach and broader government elements to take effect... The indirect approach includes empowering host nation forces, providing assistance to humanitarian agencies, and engaging key populations.[399]

Dr. Christopher Lamb, in testimony to the House Armed Services Committee in 2012, further distinguished the USSOF direct and indirect approaches by levels of cost and control.

> SOF can execute their missions directly themselves or they can conduct their missions working by, with, or through indigenous forces and populations. A number of terms have been used to describe these approaches but the terms "direct" and "indirect" are commonplace ... Often it is assumed that acting directly means employing lethal force and acting indirectly means employing non-lethal capabilites. In fact, both approaches can involve lethal and non-lethal skills. The more important differences involve costs and control. In general, acting indirectly entails lower costs but also offers less control over means employed and outcomes achieved. Acting directly can involve higher costs but

---
399 McRaven, "Posture Statement," March 06, 2012.

provides more control over the means employed and ends achieved.[400]

Dr. Lamb's distinction makes subtle observations about the trade-offs involved in USSOF power application. These distinctions are critical in application but such nuances are easily lost in the joint planning process.[401]

In 2012, US Army Special Operations Command added greater doctrinal specificity to the duality of USSOF. Special warfare and surgical strike were introduced as clear USSOF expressions of indirect and direct approaches.[402] As discussed in chapter two, "special warfare is intended to wade into uncertainty and prevail... while surgical strike is intended to squeeze out uncertainty, then execute."[403] Special warfare and surgical strike represent ARSOF's efforts to better articulate USSOF's strategic application.[404] This clarity is useful when the center of gravity method is in question. Without a clear notion of center of gravity, the direct and indirect approach paradigm becomes nebulous. The special warfare and surgical strike concepts aim to reduce this potential ambiguity. Crafting USSOF operational approaches requires connectivity to the greater joint planning logic without losing the specificity critical to understanding USSOF approaches. Special

---

400 Lamb, "Statement on 'The Future of U.S. Special Operations Forces,'" July 11, 2012.
401 Chapter one and two discusses the attempts to provide greater clarity – and public knowledge - of "special operations power." The joint planning process, at the combatant command level, relegates the implementation details of campaigns or operations to the lead service component or joint task force headquarters. This scenario is acceptable for some joint operations but it becomes problematic when steady-state Phase Zero operations are guided under many directorships, not all conversant in the varieties of power paradigms.
402 ADP 3-05, 2012.
403 Author discussion with LTG Charles T. Cleveland in December 2012, Fort Bragg, North Carolina.
404 Special warfare and surgical strike were introduced into US Army doctrine in 2012 (ADP 3-0). They are not in joint doctrine.

warfare and surgical strike present clear concepts for the use of USSOF within joint operational approaches.

Anticipation

Joint doctrine states "Joint Force Commanders must consider what might happen and look for the signs that may bring the possible event to pass."[405] Anticipation receives the shortest explanation (one half page) of the thirteen elements of design.

SOCSOUTH's operational approach exercised the art and science of anticipation. To build an operational art construct, SOCSOUTH merged two connected concepts: *assessments* and *programs*. Assessments were ongoing evaluations that appraised both Colombia capabilities and the US support to build those capabilities.[406] Programs, as detailed in chapter three, were the US-funded programs that governed the employment of US capacity in support of the Colombians. The conduct of assessments is taught in special operations qualifying courses and is routinely exercised by USSOF working with foreign partners.[407] In Colombia, assessments informed the operational and tactical decisions on where and how to apply USSOF capabilities in support of the Colombians.[408] A USSOF Advanced Operational Base (AOB) commander in Colombia linked assessments with anticipation.

> The assessments from teams on the ground were essential to understanding where to apply pressure,

---

405   JP 5-0, III-33.
406   JP 1-02, 22.
407   Author interview A19 with US Army Special Forces Colonel Paul Ott, Commander of the Special Warfare Education Group (Airborne) at Fort Bragg, NC, September 5, 2012.
408   Author interviews A02, A16, A25, A26.

transition, or push more support. Foreign internal defense across a wide country can get complex and everyone is fighting for resources. Assessments add objectivity to this process and help clear up the fog and friction. Anticipation means translating those assessments into funding and resources. This cannot be done immediately.[409]

A Special Forces officer described the relevance of assessments inside of the larger joint forces construct.

It's more important that we take the time to get the assessments done right. This may mean that we don't do something right away but it's better if we execute correctly later, affect outcomes in the right way, and get on the right side of history. The joint force mindset is on phase three [dominate] – looking at how fast we can get there. Speed is a valued quality but deterrence has a different side – a deep, contextual understanding of the problem that we are leaning into.[410]

From 1998 – 2008, USSOF in Colombia conducted a multitude of transitions, all informed by formal and informal assessments. In the uncertain environment of advising partners, assessments provide both factual content and situational awareness. USSOF assessments were both quantitative and impressionistic.[411] Written and verbal assessments were central to gauging and anticipating trends and tendencies that

---
[409] Author interview A25.
[410] Author interview with LTC Josh Walker.
[411] Author experiences in USSOF from 1997 – 2013. USSOF assessments in Colombia are classified and/or sensitive.

inform future programs.[412] The long time horizons in Phase Zero can inadvertently disconnect accurate assessments from generating the right programs. Programs are typically staffed two to four years in advance.[413] This type of anticipation is significantly different than the set-piece battle notions of anticipation. Taken in isolation, programs are a form of "hard science" in Phase Zero. Combined with assessments, programs contain a deeper philosophical element about how to leverage programs to solve problems. In Phase Zero, anticipation has a time and spatially delayed quality that spans years and even decades. COL (Ret.) David Maxwell connected assessments to the extended Phase Zero design process.

> The key in phase zero is the ability to conduct an assessment. This is one of the greatest strengths of SOF and, in particular, army special operations. Even in the absence of strategy, deployed forces working with the host nation understand the conditions, assess the environment, and can provide recommendations on what can be done … a continous presence improves the depth of these assessments … this feeds the design process with valuable data.[414]

Assessments and programs are two distinct USSOF expressions of anticipation. These elements require skilled, continuous assessments that feed the program consideration process. Ideally, this links muddy-boots actions to theater or national strategic considerations of foreign military-funded programs. Revised perspectives on anticipation can reduce

---

412 Author interviews A16, A25, A26.
413 DISAM Greenbook.
414 Author interview A05.

the gaps between the policy makers who approve the programs and the USSOF who apply those programs for strategic gain. When connected, this linkage becomes vital to achieve success in a time-extended operational art.

Operational Reach

Operational reach is "the distance and duration across which a joint force can successfully employ military capabilities ... for any given campaign or major operation, there is a finite range beyond which predominant elements of the joint force cannot prudently operate or maintain effective operations."[415]

SOCSOUTH's concept of operational reach focused on the *access* and *location* of USSOF advisors within the Colombian security infrastructure. This approach viewed operational reach as a combination of mobility platforms and the ability to leverage Colombian interior lines through host nation infrastructure.[416] Physical infrastructure included physical facilities such as airfields, roads, and fuel depots. The human infrastructure included partnered forces, local security forces, and populations.[417] Outside of major urban areas, the infrastructure within Colombian interior lines is limited. Vast jungles, high mountain regions, and weather challenges make access difficult. With its distributed small teams, USSOF operational reach combined human access and placement with the mobility platforms provided by US security assistance. By the early 2000s, SOUTHCOM's counter-drug funded security assistance programs were aggressively building Colombian

---
415 JP 5-0, III-33 to III-34.
416 Author interview A16.
417 Author interview A16.

mobility capabilities. Aviation capabilty was the cornerstone of US support. From the 1998-2008 period, the US rotary-wing program included US-built UH-1N, UH-II, and UH-60 helicopters.[418] Fixed wing and riverine assets were also included in US counterdrug assistance packages.[419] US security assistance provided advanced technical platforms (planes, ships) to transit vast geographic spaces and project Colombian combat power. In concert with the growing mobility capability, USSOF were distributed across the Colombian interior lines. The combination of capable mobility platforms and USSOF access and location provided US-Colombian partnership a formidable ability to project and sustain intelligence and security force operations.[420]

As joint doctrine implies, operational reach is very much a physics problem. In addition to mobility platforms, USSOF further expand operational reach through the access and location of USSOF advisors. This approach capitalizes on the physical and human infrastructure of the host nation. In austere environments, mobility assets may be insufficient or entirely unavailable. In such cases, alternative methods are required to extend operational reach. Proper access in the right locations are critical USSOF methods to extend operational reach.

Culmination

Culmination "is that point in time and/or space at which the operation can no longer maintain momentum."[421] Joint doctrine describes culminating points for offense (when

---

418 Ramsey III, "El Billar," 80, 108-134
419 Ramsey III, "El Billar," 80, 134.
420 Burton, "ARSOF in Colombia," 24-33.
421 JP 5-0, III-34.

continuing the attack is no longer possible), defense (when the defender can no longer shift to the offensive), and stability (when national will wanes or popular support declines).[422] Culmination ties the expenditure of resources with the timing and tempo of operations.

SOCSOUTH's campaign gauged culmination as areas where goals were accomplished and future, similar engagements would yield little benefit.[423] One example is the shift from counterdrug brigades to counterterror capabilities in the early 2000s. In foreign internal defense, culmination is closely tied to transition. When an engagement has achieved effects, scarce resources must be applied to other, higher yield requirements. SOCSOUTH, and USSOF in general, contend with another aspect related to but separate from culmination: *saturation*. Saturation involves gauging the right amount, measured in impact of presence, of USSOF advisory support to an activity or unit.[424] Saturation has an intuitive quality. It is tied to host nation perceptions about the nature, volume, and effect of US support. Saturation is gauged by a multiplicity of indicators: host nation partner receptivity, public perceptions, the tone and frequency of media coverage, and the balance of putting enough US advisory effort in place to be effective, but not more than is required. Saturated is the state "when something is unable to hold or contain more." This dictionary definition, applied to a host country, signifies a tipping point where US advisory support becomes overbearing and, thus, counterproductive. Conversely, US support can fall well below the requirements and appetite of a host nation (under saturation). In cases like Colombia, where significant

---
422 JP 5-0, III-34-35.
423 Author interview A16.
424 Author's definition. There is no joint or otherwise doctrinal definition of saturation.

US capacity is applied, over saturation is the greater risk.

Judging saturation requires a keen feel for the totality of US programs and how they impact both the population and the host government. This requires a studied, informed presence attuned to subtle indicators. Episodic engagements, and the US personnel who conduct them, are poorly suited to gauge saturation. One identifiable indicator is when consistently excellent tactical results are producing negative strategic consequences. Even combatant commanders with skilled, responsive staffs have trouble gauging saturation. Admiral (Ret.) Timothy Keating's vignette on the negative effects of US humanitarian assistance to Indonesia shows the difficulty of managing saturation. [425] USSOF assessments are critical for the continuing conversation between diplomacy, development, and defense representatives who collectively gauge and adjust saturation issues. Understanding how saturation affects USSOF assistance requires experienced judgment and a deep understanding of the host nation. In the formation of USSOF Phase Zero operational art, saturation is an essential intangible to consider in the formulation of campaigns.

Arranging Operations

Arranging Operations are the "best arrangement of joint force and component operations to conduct the assigned tasks and joint force mission. This arrangement will often be a combination of simultaneous and sequential operations ... Thinking about the best arrangement helps determine the tempo of activities in time, space and purpose."[426] Joint doc-

---

[425] Keating, Center for Strategic and International Studies (CSIS), "US Forward Presence in the Asia-Pacific Region," January 28, 2012.
[426] JP 5-0, III-35.

trine identfies specific planning tools available to arrange operations and dictate tempo. They include phases, branches and sequels, operational pause, and the implementing details included in the Time Phased Force Deployment Document (TPFDD).[427]

SOCSOUTH's campaign in Colombia used the logic and tools described in joint doctrine. There is a key difference in the method of USSOF force distribution and command and control. With small, distributed USSOF teams and with multiple orchestrating US headquarters, this construct resembles more of a network, or chain. This method combines the creative application of arranging operations with the logic of supply chain management.[428] This hybrid is expressed as an *arranging chain*. In Colombia, arrangement of USSOF operations occured at multiple levels by small, forward-stationed teams. SOC-Forward remained the overall synchronizer. But the improvisational and distributed qualities of such an advisory effort conceded a measure of control to small teams forward with the Colombians. USSOF teams exerted influence through their advisory actions both geographically and functionally. For example, military information support teams (MISTs), civil affairs teams, and US Army, Navy, and Air Force special operations teams conducted simultaneous and, at times, sequential support to Colombian operations. Within these formations, small units exercised a distribution (materiel) and influence (ideas) chain to purposefully connect tactical

---

[427] JP 5-0, III-35-36.
[428] The Association for Operations Management (APICS), APICS Dictionary, 13th ed. Online. Supply chain management is the "design, planning, execution, control, and monitoring of supply chain activities with the objective of creating net value, building a competitive infrastructure, leveraging worldwide logistics, synchronizing supply with demand and measuring performance globally."
http://www.apics.org/gsa-main-search?q=#supply%20chain%20managment|dictionary (accessed February 1, 2013).

actions to strategic outcomes. A Special Forces Advanced Operational Base commander stated,

> The indirect methods of military information support and civil affairs are influencing the population with specific themes and ideas. The information operations are so valuable in adapting quickly to local conditions. The [Special Forces Advanced Operational Base] is driving bigger strategic objectives through small actions, day in and day out, that are hard to template. But when you see the opportunities, you go for them… The Colombians are leading and we are supporting, so don't always expect an OPORDER brief. The easy part is the tactical training – the tough part is connecting all this, in short time frames, to the larger strategy.[429]

The arranging chain also describes how operational art is formulated without a unilateral command synchronizing all actions. Optimally, operational art is conceptualized and directed by a single operational artist: the commander executing the campaign. Phase Zero does not afford this luxury of unity of command and centralized synchronization. In Phase Zero, operational art is conducted across an extended network, or chain of commanders and influencers (Figure 15).

---

[429] Author interview A11. This special forces officer worked command and staff positions in Latin America, to include Colombia, from 2001-2010. OPORD is shorthand for an operations order. An operations order is a detailed plan that is distributed both higher and lower to coordinate all military actions. A Special Forces Advanced Operational Base (AOB) is a headquarters element of between twelve and forty personnel commanded by a Major or Lieutenant Commander (O-4) level officer.

ANALYSIS OF SPECIAL OPERATIONS PHASE ZERO OPERATIONAL ART

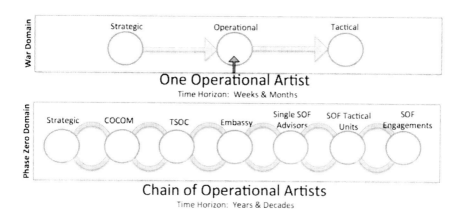

Figure 15. Arranging Chain

This method of operational distribution creates a unique style of command and control. Within this organizational architecture, operational art is a distributed endeavor. The disadvantages are the multiple points of failure, different interpretations of success, and, over time, there can be a loss of the long-term vision. This distributed method makes the construct of a coherent operational art very fragile. The advantage of an arranging chain of operational artists is the value of a low visibility presence, a nimble posture suited for improvisation, and a rapid execution cycle. These advantages allow subordinate elements to seize fleeting opportunities that would otherwise go unrecognized by a higher level joint force headquarters. This command and control arrangement approximates the diplomatic dynamic where "different actors dominate in different cycles."[430] This arrangement also favors interagency collaboration which is less concerned with hierarchical chains and more interested in direct access at the

---

[430] James Scott, "Interbranch Rivalry and the Reagan Doctrine in Nicaragua," *Political Science Quarterly* 112, Number 2, 1997, 237-260.

point of influence. As with other USSOF expressions of design, the arranging chain places a premier on posture. Where host nations dictate the operational plans and pace, USSOF advisory efforts must adapt its own style of operational artistry. The arranging chain is one such adapation that postures USSOF to construct unique operational art approaches.

Summary

To craft an operational approach to a problem, joint doctrine provides thirteen elements of operational design. Four of these elements apply to special operations with few, if any, distinctions. They are *objectives, effects, lines of operation / lines of effort,* and *forces and functions*. The remaining nine elements have special operations expressions that deviate from the doctrinal elements. These distinctions are revealed by extracting special operations elements of design from their closest joint doctrine relative. By juxtaposing the special operations modification from its parallel joint doctrine term, a question is formed. Is the special operations element of design a replacement for, modification to, or expression of, the joint doctrinal element? From this answer, a logic begins to form. From this logic, the special operations operational approach starts to take shape.

Where a replacement is recommended, the doctrine steers thinking in the wrong direction. Where a modification is recommended, the doctrine simply needs an adjusted waypoint to guide planning logic. Where USSOF expressions are recommended, then the doctrine peformed its function adequately, but not perfectly; that is, the logic remains valid but the SOF expression of such logic requires clarification. Certain USSOF expressions may require contextual explanation because

ANALYSIS OF SPECIAL OPERATIONS PHASE ZERO OPERATIONAL ART

many actions and activities are outside of normal military education curricula.[431]

In devising USSOF Phase Zero operational approaches, two joint terms are recommended for replacement. *Termination* is replaced by *transition* and *military end-state* is replaced by *position of continuing advantage*. Two terms are recommended for modification. *Culmination* is modified as *saturation*. *Center of gravity* is modified as *right partner, right place, right time* (R3). To clarify, R3 is an inadequate substitute for the analytical benefits afforded by a doctrinal center of gravity analysis. Where R3 is useful is in steering the arrangement of posture, operations and activities. The final six elements are classified as USSOF expressions of the design elements. *Decisive points* have an expression of *decisive relationships*. *Direct and indirect approach* are expressed as *surgical strike and special warfare*. *Anticipation* is expressed by both *assessments and programs*. *Operational reach* has a unique expression in *access and location*. Finally, *arranging operations* are better understood by the concept of an *arranging chain*.

These recommendations have implications that go beyond joint military doctrine. Chapter six offers conclusions and implications that span from the lowest tactical level to the higher realms of strategy.

---

[431] Special operations and special operations forces have a tenuous footing in professional military education (PME) realms. For example, in the US Army, there are few codified special operations requirements in education until the Command and General Staff College (CGSC) level. CGSC is normally attended by an officer in his or her tenth to twelfth year of service. In the 10.5 month course, there are four hours devoted to special operations. Most officers gain knowledge of special operations and special operations force through experience and not education. (Reference: Author experience as Special Operations Leadership Development & Education Chief at the US Army Command and General Staff College, May 2011 to May 2012.)

6
Conclusions

Six conclusions are drawn from examining special operations approaches in Yemen, Indonesia, Thailand, and Colombia. This study included over thirty interviews with Phase Zero practitioners, many with decades of experience. Anecdotally, the below conclusions also represent special operations observations from other regions, most notably Africa. Chapter six presents six conclusions followed by suggestions for further study and recommendations. Finally, the implications of improved engagement strategies are discussed.

Defining the special operations operational art for Phase Zero is not a straightforward proposition. This study outlined the contours of special operations campaigns, or campaign aspirants, in Yemen, Indonesia, Thailand, and Colombia. The evidence suggests that USSOF is successfully crafting new campaign modes guided by revised operational art constructs. From this evidence, six conclusions were drawn:

- special operations *do* execute distinct Phase Zero campaigns propelled by innovative operational art
- the US DoD joint doctrine elements of operational

CONCLUSIONS

design require modifications to better guide USSOF Phase Zero campaign planning
- USSOF campaigns contain logic that is often paradoxical to the generally applied principles of military force
- Phase Zero presents a particularly difficult environment to formulate and apply an operational art
- Phase Zero operational art executed by USSOF combines supply chain management-style structure with network logic; the hybrid is a type of "arranging chain" that makes operational artistry possible over vast time, distance, cultural, and programmatic spans (Figure 1, 14)
- despite their clear distinctions, USSOF campaigns are inextricably tied to and in support of the macro strategy of the combatant commander and the US country team.

The first conclusion supports the thesis of this study. **Special operations Phase Zero campaigns are a nuanced application of specialized forces, often in domain gaps, that recognize diplomacy as the leading art.** These campaigns view Phase Zero as a singular phase unto itself, *not* just in preparation for potential larger, intrusive military operations. The sensitivities of Phase Zero require adjustments in environmental understanding, foreign policy aims, and the use made of US military power. To succeed in this environment, special operations approaches exhibit four consistent characteristics. (1) USSOF, by its quality and design, employs distributed, low-visibility methods to achieve foreign policy objectives in remote, unstable regions. Degradation of the adversary is a critical but often temporal accomplishment. Lasting progress is achieved by applying the right level and quality of US influence with host nation counterpart agencies. Impact is measured by increased host nation capability, positive public perceptions of host nation actions, and relationships that can survive crises. (2) At the ground level, USSOF campaigns are woven from disparate programs, authorities, and agreements executed by small, culturally conversant teams. When crafted and executed properly, USSOF campaigns are smaller in scale but not in impact. Like the policies they support, these approaches are applied incrementally. Likewise, results are gradual. With few exceptions, Phase Zero is not the realm of decisive action. (3) If a strategic surprise requires decisive military actions, USSOF are uniquely capable of shifting to a lead or supporting role. (4) USSOF campaigns are interagency interdependent. As part of the interagency team, special operations are brokers of knowledge that fulfill a clear niche: connecting field-level persistent engagements to the attainment of strategic objectives. Taken together, these characteristics typify the distinct application of special operations power.

Second, the joint doctrine elements of operational design (Joint Publication 5-0, Joint Operation Planning) require modifications to better guide USSOF Phase Zero campaign planning. Nine of the thirteen elements, conceived for use in joint warfare, show reduced utility in a Phase Zero campaign. Though there is great flexibility in joint doctrine planning constructs, a modified planning template improves the cognitive linkage of special operations engagement methods to strategic objectives. This study recommends modifications to nine of the thirteen elements of operational design (Figure 14, SO Phase Zero Expressions). These modifications better guide USSOF campaign planning and can assist non-SOF (general purpose forces, US State Department, policy-makers, non-governmental organizations) in understanding the organizing logic and application of special operations (and other DoD actions) in Phase Zero. Advantages are won by aligning design terminology closer to the diplomacy-centric essense of Phase Zero. Intentions are expressed more clearly when mismatched terminology is avoided. Frictions among SOF, DoD, DoS, policy makers, and NGOs will remain. But when commonality of purpose is easier to ascertain, frictions are reduced.

The third conclusion recognizes five paradoxes in the logic of special operations Phase Zero campaigns. These apparently self-contradictory truths may appear contrary to standard, sound tactical wisdom.[432] The five paradoxes often evident are:

- less is better than more
- steady and slow is (often) preferred over intrusive and fast
- a supporting role is better than a lead role
- the wrong man can do more harm than the right man can do good[433]
- conceding military control and precision can create better long-term outcomes

---

[432] Luttwak, *The Logic of War and Peace*, 1-15. Luttwak provides a book-length study on the paradoxical nature of strategy. The above conclusion is influenced by Luttwak's thesis that "the entire realm of strategy is pervaded by a paradoxical logic very different from the ordinary 'linear' logic by which we live in all other spheres," 2. The paradoxes referenced here are oriented on the widely applied logic of the use of overwhelming military force. Called by many the "American way of war" or the Weinberger/Powell doctrine, the US military remains philosophically tied to the use of overwhelming force. This breeds a tactical mindset that discarding clear US (asymmetric) advantages adds unnecessary risk. This sound tactical premise becomes problematic when applied to strategic environments, particularly in Phase Zero.

[433] Major Fernando M. Lujan, USA, "Light Footprints: The Future of American Military Intervention," Center for New American Studies (CNAS), March 2013. http://www.cnas.org/files/documents/publications/CNAS_LightFootprint_VoicesFromTheField_Lujan.pdf (accessed March 27, 2013). Lujan credits this aphorism to The Office of Strategic Services (OSS) Assessment Staff, Assessment of Men: Selection of Personnel for the Office of Strategic Services, (New York: Rinehart and Co., 1948).

A fifth principle is not paradoxical, but is no less important: personal (relational) determines the outcome of the programmatic (transactional). These paradoxes are not absolutes, nor are they without their own risks. An historical pattern is that these principles, and the actions that they drive, become disconnected over time.[434] The challenge is *quality* of personnel and *continuity* of execution.[435] Because these campaigns are stretched over years, successive decision-makers and their staffs must understand past, current, and proposed actions. Their contextual understanding is challenged by extended timelines, remote regions, and cultural peculiarities. Further complicating this problem is the confounding logic typified by the five paradoxes. In this situation, a well-conceived but fragile operational art rhythm is at risk. A further pitfall awaits: whereas the "right man" can do so much in these environments, it takes just one "wrong man" in the so-called arranging chain to disrupt a hard-won trust. Individuals and small teams lack redundancy and rely greatly on experienced personnel with exceptional professional qualities; it is a persistent challenge to assess, select, educate, and train the right personnel with the skills and maturity to succeed in this environment.

Improved elements of operational design increase the probability that such paradoxes become understood and embraced. By reducing the conflict between linear logic and

---

[434] Internal USSOF rotations can further disrupt continuity. When operating under very broad guidance, rotations between small teams can bring large swings in focus and emphasis. For example, one team might focus six months of effort on host nation influence operations through radio, personal engagement and deception. The next team might come in and determine that kinetic actions are more important. A strong host nation command and institutions can prevent this disruption to a certain extent.

[435] The continuity challenge has two sides: strategic and programmatic. Strategic continuity is difficult given the cyclic nature of US elections and their impact on long-term strategic thinking. Program continuity is challenged by the simple premise that a rotation of personnel acting on a single project introduces swings in competence, emphasis, and understanding.

Phase Zero (paradoxical) logic, unity of purpose is more easily accomplished. In a final paradoxical twist, "victory" in the realm of Phase Zero resembles more of a whimper than a bang. As in the Colombia case study, success allows a measured reduction in programs, capacity, and involvement, with little fanfare attached and no true decisive finality. Successful Phase Zero outcomes, like strategy itself, are temporal in nature and may not generate a visible political success nor generate public recognition.

Fourth, Phase Zero presents a grim challenge for the formulation of operational art. US Phase Zero efforts are led by policy and diplomacy. In this environment, military operational art is necessarily a subordinate consideration. With military actions restrained and purposefully limited, the free hand of the operational artist is constricted. The sovereign host nation introduces yet another set of limitations, both specified and implied. To further complicate matters, the US may lack a clear strategy. With one or all of these factors in play, even an exquisite concept of USSOF operational art is unlikely to produce lasting strategic outcomes. Overall, the Phase Zero environment is unfriendly to the construction and application of operational art.

But as this study shows, there are ways to work through these challenges. In SOCSOUTH, the hostage rescue mission galvanized the US effort and lent strategic clarity to the US mission in Colombia. The crisis gave greater urgency to the joint US-Colombian effort. But it also provided a clearly visible manner in which tactics could link to strategy. Conversely, in Thailand, there is little strategic urgency to move beyond a balanced, near-persistent USSOF engagement strategy. With less ambitious strategic aspirations, the special operations Phase Zero operations, actions, and activities need only form a nascent campaign. There is value in limitations. Tactical overreach in Thailand would likely result in counterproductive strategic outcomes.

Common to all scenarios is that operational art cannot form when US strategy is adrift. In these cases, USSOF Phase Zero operations conduct precursor activities in advance of potential foreign policy options. USSOF assess, understand, build relationships, and improve host nation capability or capacity. This lightly footprinted, small investment is a visible sign of US policy-in-action and provides a hedge against strategic surprise. Potential foreign policy options might include supporting a government (foreign internal defense), overthrowing or disrupting a government (unconventional warfare), or singular actions outside of an operational art (hostage rescue).

Fifth, operational art executed by special operations in Phase Zero combines supply chain management principles with network logic; its hybrid is an "arranging chain" that makes operational artistry possible over vast time, distance, cultural, and programmatic spans. Phase Zero presents a wicked environment that requires the sophisticated logistical agility of supply-chain management with an agile, pseudo-hierarchical, and networked mindset. (Figure 1, 14). In this environment, the notion of a single, commander-like operational artist synchronizing all events in time, space, and purpose is invalid. The arranging chain describes a USSOF posture of connected, distributed nodes that are hierarchical for control and resources yet are networked for ideas and opportunities. The abiding premise is that different actors (nodes) dominate in different cycles, thus creating a jazz-style, "improvisation on a theme" implementation of operational artistry.[436] This

---

[436] Vali Nasr, *The Dispensable Nation: American Foreign Policy in Retreat* (New York: Doubleday, 2013), 49. Nasr cites this quote as a favorite of former US Ambassador

method adds risk by conceding and dispersing control yet it capitalizes on the ability to influence consistently and persistently across the vast physical and human infrastructure known as the "Global SOF Network."[437]

The final conclusion is that USSOF campaigns are tied to and in support of the overall strategy of the combatant commander and the US country team. This study deliberately isolated USSOF campaigns in order to ascertain their unique characteristics and logic. This should in no way be interpreted that USSOF campaigns are separate from the directed strategies and guiding campaign plans of the combatant commander. USSOF campaigns are regional or country-level implementations of the combatant commander's Theater Campaign Plan. USSOF campaigns are further guided by USSOCOM's global combating terrorism responsibilites. Yet another level of direction comes from within the country at the behest of the US Ambassador. At the level of execution, the host nation adds even more complexity with its national challenges and aspirations. In this environment, context is paramount. The formulation of Phase Zero campaigns, at the regional and country level, is an art that is formulated and arranged below the theater commander level.[438] In this complex environment, USSOF are one tool available to combatant

---

Richard Holbrooke.
437 Admiral William H. McRaven, USSOCOM Posture Statement before the 113th Congress, Senate Armed Services Committee, March 5, 2013.
438 As discussed in chapter two, combatant command Theater Campaign Plans (TCP) are the overarching, organizing campaign plans for the four-star theater commander. Generally, combatant commands are capable of synchronizing a few, select regions or campaigns. The other campaigns, or campaign-like arrangements are synchronized by lead services or sub-unified commands. For example, US Pacific Command has over 40 countries in its area of responsibility. It would be disingenuous to claim that the combatant commander is the operational artist for the theater, for all the regional constructs, and for all the countries. While combatant commanders set the priorities and authorize and employ the forces and resources, the arrangement of actions is done by smaller, joint or service-led headquarters, particularly at the country level.

commanders and ambassadors to coordinate and arrange US interests, either as lead or in support. USSOF are missioned, manned, and trained to provide distinct options up to, and including, full campaigns. A key challenge in applying special operations against combatant command problems is integrating potentially divergent philosophies on the use of military power. Revised concepts that better link USSOF logic and approaches to joint force constructs are one way to address this tension and improve joint force effectiveness.

Areas for Further Study and Recommendations

Two compelling concerns merit further exploration and study. First, doctrine modifications have far-reaching impacts. Seemingly innocuous tweaks of language in doctrine are akin to shifting the foundation of a house. All vertical and horizontal concepts are affected. Doctrinal shifts generate reconsiderations of everything from warfighting methods to budgetary allocation. Doctrine is not dogmatic but it is authoritative.[439] Doctrine changes risk disrupting the balanced logic of sound US military capabilities, principles, and application. For these reasons, doctrine changes are difficult and slow. A negative byproduct is that doctrine is slow to respond when the environment demands new approaches. To implement the recommendations of this study, further examination is required to determine the implications for substantive changes in joint doctrine.

Second, special operations education, and by extension all military professional education, must be evaluated. Internally, USSOF have a growing, but relatively small, strategic culture. Crafting an operational art in Phase Zero requires a broad

---
[439] JP 1-02, 90.

and deep USSOF education and experience base. To reach their strategic potential, the USSOF community requires a more coherent education program on the history, theory, doctrine, and practice of operational art. Like many military forces, tactical acumen is the first, and sometimes the only, area where mastery is achieved. This is necessary but not sufficient. USSOF need greater mastery in the area of policy and theater-level strategy to consistently develop feasible operational art. Moreover, USSOF need envoys in the military, government, and civilian sectors who can articulate and, when appropriate, advocate for such options. To accomplish this, more analysis is needed. In the specific realm of operational art, the most pressing area for further study is how *special warfare* and *surgical strike* are combined to create fungible applications of operational art.

External to USSOF, greater education is needed regarding the strategic application of special operations. Too often, the commando-style exploits of special operations are the prevailing consideration for USSOF employment. In other cases, the secrecy of special operations discourages investigation, particularly from academia. Outside of the special operations community, a deeper understanding is required on the capabilities, limitations, and strategic utility of special operations. The strategic use of special operations needs greater exposure in military professional military education (PME) venues, within the interagency, and in foreign policy curricula. Without this understanding, a gap will remain between the formulation of strategy and policy, and the "left of war" options offered by special operations.

To start, one recommendation can be implemented now: a renewed educational emphasis exploring special operations expressions of operational art. This starts within the special operations community but is certainly appropriate for the joint force. With special operations activities occuring in over 75 countries, there is a rich body of work to examine. From there, the special operations community must educate its own force with a well-burnished articulation of operational art. Important in this internal USSOF dialogue is the input of the US joint force and interagency partners. Exquisitely formed USSOF operational art that is too detached from US joint doctrine or too opaque for the interagency is useless. The application of operational art by special operations in Phase Zero may gain deep interest only from a limited strategic caste. However, this does not limit the export of broadly understood principles that can be readily understood and applied by security and policy professionals, writ large.

Implications

This study began with the observation that the US lacks a grand strategy. Without a monolithic threat to defeat, the US employs engagements to pursue US interests abroad. Defined as "the active participation of the United States in relationships beyond our borders," engagements themselves are not a substitute for strategy.[440] Engagements, appropriately conceived and properly arranged, realize specific policy goals or contribute to vague strategic notions. But with a clear strategic purpose, a consistent resource commitment, and a sound grasp of appropriate power principles, engagements can form operational art. Consequently, operational art ensures that engagements become a more effective means to achieve US strategic ends.

A Phase Zero operational art creatively crafts engagements

---
440 National Security Strategy, 2010, 19.

to achieve strategic ends *under* the threshold of war. In politically sensitive environments, special operations offer a distinct application of US power and influence. Revised operational constructs are required to express how special operations provide unique policy options in complex environments, in domain gaps, and where war-like actions or a large US presence is ill-advised. To do this, the US requires amended concepts of operational artistry where special operations are considered as the best option.

This study explored how special operations, in conjunction with other forms of influence, can apply operational art to skillfully achieve US strategic objectives. Where large-force military actions are potentially counterproductive, the application of a creative, non-standard operational art construct has tremendous strategic potential. With a reduced USSOF commitment to major land wars, the US gains a strategic dividend for Phase Zero: an extraordinary special operations capability within the most capable joint force in history. This dividend combined with an improved application of operational art in Phase Zero offers that which is most prized by strategists and policy makers: increased strategic options for the protection and pursuit of US interests. The implications go beyond the "going small" mantra of foreign policy advocates who favor lightfooted approaches. The emerging application of operational art by special operations in Phase Zero is about going smart, going long, and going local, *while* going small. Future US strategy – perhaps even a grand strategy - will require revised and innovative applications of US power, smaller in scale but not in effect. Paradoxically, going big might just require the art of getting small.

List of Abbreviations

ADM	Army Design Methodology
ADP	Army Doctrinal Publication
AECA	Arms Export Control Act
AFSOC	Air Force Special Operations Command
ALP	Afghan Local Police
AOB	Advanced Operational Base
AQAP	Al Qaeda in the Arabian Penninsula
ARSOF	Army Special Operations Forces
AUC	Autodefensas Unidas de Colombia
BRCNA	Brigade Contra el Narcotrafica
CCOPE	Commando Conjunto de Operaciones Especiales
CERTE	Colombian Army Tactical Retraining Center
CNP	Colombian National Police
COCOM	Combatant Command
COESE	Commando de Operaciones Especiales Del Ejercito

LIST OF ABBREVIATIONS

COG	Center of Gravity
COIN	Counterinsurgency
CST	Cultural Support Team
CT	Counterterrorism
CTFP	Combatting Terrorism Fellowship Program
DOD	Department of Defense
DOS	Department of State
ELN	Ejercito de Liberacion Nacional
FAA	Foreign Assistance Act
FARC	Fuerzas Armadas Revolucionares de Colombia
FID	Foreign Internal Defense
F3EAD	Find, Fix, Finish, Exploit, Analyze, Disseminate
GCC	Geographic Combatant Command
GWOT	Global War on Terror
IDAD	Internal Defense and Development
IMET	International Military Exchange Training
INL	International Narcotics and Law Enforcement Affairs
JP	Joint Publication
JSOTF	Joint Special Operations Task Force
JSOU	Joint Special Operations University
JTF	Joint Task Force

JUSMAGTHAI	Joint United States Military Advisory Group Thailand
KOPASUS	Indonesian Special Forces Command
MARSOC	Marine Corps Forces Special Operations Command
MISO	Military Information Support Operations
NATO	North Atlantic Treaty Organization
NAVSPECWARCOM	Naval Special Warfare Command
NSS	National Security Strategy
OEF-A	Operation Enduring Freedom – Afghanistan
OIF	Operation Iraqi Freedom
OPCON	Operational Control
PATT	Planning and Assistance Training Teams
RTA	Royal Thai Army
RTA SWCOM	Royal Thai Army Special Warfare Command
R3	Right Partner, Right Place, Right Time
SAMS	School of Advanced Military Studies
SO	Special Operations
SOC	Special Operations Command
SOF	Special Operation Forces
SOJTF	Special Operations Joint Task Force

LIST OF ABBREVIATIONS

TCP	Theater Campaign Plan
TSOC	Theater Special Operations Command
UCP	Unified Command Plan
USAID	United States Agency for International Development
US	United States
USASFC	United States Army Special Forces Command
USASOC	United States Army Special Operations Command
USSF	United States Army Special Forces
USSOCOM	United States Special Operations Command
USSOF	United States Army Special Operations Forces
UW	Unconventional Warfare
VSO-ALP	Village Stability Operations – Afghan Local Police
WWII	World War Two
3D	Diplomacy, Defense, Development

Bibliography

Primary Sources

Interviews

A08, US Army Special Forces Lieutenant Colonel. Interviewed by author. Fort Leavenworth, Kansas. 6 August 2012.

A17, US Army Special Forces Major. Interviewed by author. Fort Leavenworth, Kansas. 23 August 2012.

A18, US Army Special Forces Major. Interviewed by author. Fort Leavenworth, Kansas. 23 August 2012.

A20, US Army Special Forces Lieutenant Colonel. Interviewed by author. Fort Leavenworth, Kansas. 20 August 2012.

A21, Retired US Army Special Forces Colonel Interviewed by author. Fort Leavenworth, Kansas. 8 September 2012.

A25, US Army Special Forces Lieutenant Colonel. Interviewed by author. Fort Leavenworth, Kansas. 25 September 2012.

A26, US Army Special Forces Major. Interviewed by author. Fort Leavenworth, Kansas. 6 September 2012.

A30, US Army Special Forces Lieutenant Colonel. Interviewed by author. Fort Leavenworth, Kansas. 24 September 2012.

A31, US Department of State official. Interviewed by author. Fort Leavenworth, Kansas. 25 September 2012.

Ames, Russ, Major. Interviewed by author. Fort Leavenworth, Kansas. 6 September 2012.

Ancker, Clint. Interviewed by author. Fort Leavenworth, Kansas. 29 August 2012.

Benson, Kevin, Colonel (USA Ret.). Interviewed by author. Fort Leavenworth, Kansas. 20 August 2012.

Burns, Barrett, Lieutenant Colonel. Interviewed by author. Fort Leavenworth, Kansas. 16 July 2012.

Celeski, Joseph Colonel (USA Ret.). Informal discussions with author, March 2012.

Depolo, Michael, Lieutenant Colonel. Interviewed by author. Fort Leavenworth, Kansas. 21 August 2012.

Donahoe, Adrian, Lieutenant Colonel. Interviewed by author. Fort Leavenworth, Kansas. 20 August 2012.

Fontenot, Gregory, Colonel (USA Ret.). Interviewed by author. Fort Leavenworth, Kansas. 20 August 2012.

Gleiman, Jan Kenneth, Lieutenant Colonel. Interviewed by author. Fort Leavenworth, Kansas. 10 August 2012.

Greer, James, Colonel (USA Ret.). Interviewed by author. Fort Leavenworth, Kansas. 6 September 2012.

Kenny, Michael, Lieutenant Colonel. Interviewed by author. Fort Leavenworth, Kansas. 6 August 2012.

Maxwell, David, Colonel (USA Ret.). Interviewed by author. Fort Leavenworth, Kansas. 30 July 2012.

Ott, Paul, Colonel. Interviewed by author. Fort Leavenworth, Kansas. 5 September 2012.

Rainey, James, Colonel. Interviewed by author. Fort Leavenworth, Kansas. 29 August 2012.

Riley, Paul, Colonel. Interviewed by author. Fort Leavenworth, Kansas. 21 August 2012.

Rogan, Michael, Chief Warrant Officer, Four (CW4). Interviewed by author. Fort Leavenworth, Kansas. 18 July 2012.

Rosengard, Mark, Colonel (USA Ret.). Interviewed by author.

Fort Leavenworth, Kansas. 8 September 2012.

Schnell, Steven, Major. Interviewed by author. Fort Leavenworth, Kansas. 13 August 2012.

Skoric, Dale, USAID Representative. Interviewed by author. Fort Leavenworth, Kansas. 13 August 2012.

Sepp, Kalev (Ph.D). Interviewed by author. Fort Leavenworth, Kansas. 22 August 2012.

Walker, Joshua, Lieutenant Colonel. Interviewed by author. Fort Leavenworth, Kansas. 18 July 2012.

Wilson, Greg, Colonel. Interviewed by author. Fort Leavenworth, Kansas. 23 August 2012.

Written Reports or Statements

Beittel, June S. *Colombia: Background, US Relations, and Congressional Interest.* Washington DC: Congressional Research Service, 28 November 2012.

Berridge, Geoff. Interview by Georgie Day. "Geoff Berridge on why we need diplomats." *The Browser: Writing worth reading* (unpublished date). http://thebrowser.com/interviews/ geoff-berridge-on-why-we-need-diplomats (accessed 12 December 2012).

_____. *Diplomacy: Theory and Practice.* Hertfordshire, UK: Prentice Hall, 1995.

Celeski, Joseph D, Colonel (USA Ret.), *Political Warfare*, draft manuscript, October 2012.

Celeski, Joseph D, Colonel (USA Ret.), Slemp, Timothy S., Lieutenant Colonel (USA Ret.), Jogerst, John D., Colonel (USAF Ret.). *An Introduction to Special Operations Power: Origins, Concept, and Application.* Draft manuscript, April 2013.

Chanlett-Avery, Emma Dolven, and Ben Dolven. *Thailand:*

Background and US Relations. Report, Washington DC: Congressional Research Service, 5 June 2012.

Clinton, Secretary of State Hillary Rodham. "Remarks at the Special Operations Command Gala Dinner, Tampa, FL." Washington, DC, 23 May 2012. http://still4hill.com/2012/05/24/ hillary-clinton-at-tampa-socom-gala/ (accessed on 15 December 2012).

Cordesman, Anthony H. "Department of Defense, State Department, USAID and NSC Reporting on the Afghan War." Center for Strategic Studies and International Studies (CSIS), 19 May 2010. http://csis.org/publication/department-defense-state-department-usaid-and-nsc-reporting-afghan-war (accessed 14 January 2013).

Craddock, Bantz J. *Posture Statement of General Bantz J. Craddock*, Commander, US Southern Command, before the 109th Congress House Armed Services Committee, 16 March 2006.

Defense Institute of Security Assistance Management (DISAM). *The Management of Security Cooperation (Greenbook)*. 31st edition. Washington DC: Government Printing Office, February 2012.

Department of the Army. *Army Activities in Underdeveloped Areas Short of Declared War*. Memorandum For the Secretary of the Army, Washington DC: Brigadier General Richard G. Stilwell, 13 October 1961.

Department of Defense. *Theater Campaign Planning: Planner's Handbook*, Version 1.0. Office of the Deputy Assistant Secretary of Defense for Plans, Officer of the Undersecretary of Defense for Policy. Washington DC: Government Printing Office, February 2012.

Department of State. *13 Dimensions of a Foreign Service Officer*. http://careers.state.gov/ resources/downloads/

downloads/13-dimensions (accessed 15 January 2013).

———. *2010 Human Rights Report: Thailand*. Country reports on human rights practices, 8 April 2011. http://www.state.gov/j/drl/rls/hrrpt/2010/eap/154403.htm (accessed 21 December 2012).

———. *Interagency Conflict Assessment Framework*, July 2008, http://www.state.gov/documents/organization/187786.pdf (accessed 14 January 2013).

Department of State Accountability Review Board (ARB). *Unclassified report investigating the 11-12-2012 attack on the U.S. Consulate in Benghazi*. http://www.scribd.com/doc/118488083/State-Department-Investigation-of-Benghazi-Attack-of-9-11-2012 (accessed 28 January 2013).

Feith, Douglas J. "Transformation and Security Cooperation, remarks by Under Secretary of Defense for Policy Douglas J. Feith." Washington DC, 8 September 2004. http://defenselink.mil/Speeches/Speech.aspx?SpeechID=145 (accessed 12 November 2012).

Fraser, General Douglas M. *Posture Statement of General Douglas M. Fraser, United States Air Force, Commander, United States Southern Command*, before the 112th Congress House Armed Services Committee, 6 March 2012.

Hill, General James T. "Statement before the House Armed Services Committee on the State of Special Operations Forces," 12 March 2003. http://armedservices.house.gov/comdocs/ openingstatementandpressreleases/108th congress/03-03-12hill.html (accessed 17 January 2013).

Holloway III, Admiral James L. *Iran Hostage Rescue Mission Report (Unclassified)*. Special Operations Review Group, Washington DC: The Navy Department Library, 23 August 1980. http://www.history.navy.mil/library/online/hollowayrpt.htm (accessed 14 January 2013).

Human Rights Watch. *World Report 2011: Thailand*. http://www.hrw.org/world-report-2011/thailand (accessed 21 December 2012).

Lamb, Christopher J. "Statement of Christopher J. Lamb, Distinguished Research Fellow, Center for Strategic Research, Institute for National Strategic Studies, National Defense University on 'The Future of U.S. Special Operations Forces'." Testimony before the Subcommittee on Emerging Threats and Capabilities, House Armed Services Committee, U.S. House of Representatives, 11 July 2012.

Lawson, Marian Leonardo, Susan B. Epstein, and Tamara J. Resler. *State, Foreign Operations, and Related Programs: FY 2011 Budget and Appropriations*. Report, Congressional Research Service, 22 April 2011. http://www.fas/org/sgp/crs/row/R41228.pdf (accessed 31 January 2013).

McRaven, Admiral William. "Commander's Address to Sovereign Challenge IX, June 6, 2012." San Jose, CA: U.S. Special Operations Command, 2012.

_____. Posture Statement, Admiral William H. McRaven, USN, Commander, United States Special Operations Command, before the 112th Congress, United States Senate, March 06, 2012. Tampa, FL: http://www.socom.mil/Documents/ 2012_SOCOM_POSTURE_STATEMENT.pdf (accessed 23 September 2012).

_____. Posture Statement, Admiral William H. McRaven, USN, Commander, United States Special Operations Command, before the 113th Congress, United States Senate, March 05, 2013. Tampa, FL: http://www.armed-services.senate.gov/statemnt/2013/03%20March/McRaven%2003-05-13.pdf (accessed 15 April 2012).

Obama, Barak. "Presidential Letter - 2012 War Powers

Resolution 6-month Report." The White House Office of the Press Secretary, Washington DC, June 15, 2012. http://www.whitehouse.gov/the-press-office/2012/06/15/presidential-letter-2012-war-powers-resolution-6-month-report (accessed 18 December 2012).

Pastrana, Andres. "Plan Colombia: Plan for Peace, Prosperity, and Strengthening the State." Office of the Presidency, Government of Colombia, October 1999. http://www.presidencia.gov.co./webpress/plancolo/plancin2.htm (accessed 17 January 2013).

Sacolick, Major General Bennet S. and Grigsby, Brigadier General Wayne W., Jr. "Special Operations/Conventional Force Interdependence: A Critical Role in 'Prevent, Shape, Win'." *Army Magazine* 62, no. 6 (June 2012): 39-42. http://www.ausa.org/publications/ armymagazine/archive/2012/06/Pages/default.aspx (accessed 3 October 2012).

Salaam-Blyther, Tiaji. *USAID Global Health Programs FY2001-FY2012 Request*. Washington, DC: Congressional Research Service, 30 June 2011.

Serafino, Nina M. *Colombia: The Uribe Administration and Congressional Concerns*. Washington, DC: Congressional Research Service, 14 June 2002.

Sharp, Jeremy M. *Yemen: Background and US Relations*. Report, Washington DC: , Congressional Research Service, 12 November 2012. http://www.fas.org/sgp/crs/mideast/RL34170.pdf (accessed 15 December 2012).

The Constitution of the United States. Washington, DC: US National Archives, 1789.

The White House. *A National Security Strategy for a New Century*. Washington, DC: The White House, 1998.

———. "A National Security Strategy for Engagement and

Enlargement." *National Security Strategy*, Washington DC: The White House, February 1995, 33.

_____. *National Security Strategy of the United States of America, May 2010* (Washington, DC: The White House, 2010). http://www.whitehouse.gov/sites/default/files/rss_viewer/national_security_strategy.pdf (accessed 26 September 2012).

_____. *National Security Strategy of the United States of America*, January 1993. Washington DC: The White House, 1993.

_____. *The National Security of The United States of America*. Washington, DC: The White House, 2006.

_____. *The National Security Strategy of the United States of America*. Washington, DC: The White House, 2002.

United Kingdom Ministry of Defence. "The British Army Home Page," 19 October 2010. http://www.army.mod.uk/news/24264.aspx (accessed 26 September 2012).

United States Agency for International Development (USAID). "3D Planning Guide: Diplomacy, Defense, Development (15 AUG 2011 Draft)," August 15, 2011.

_____. "Civil Military Operations Guide, Version 2.3 (Draft)." Washington DC, February 2012.

United States Army Special Operations Command (USASOC), unclassified command brief (power point), June 2012.

United States Army War College. *Campaign Planning Handbook*. Carlisle Barracks, PA: Department of Military Strategy, Planning and Operations, Academic Year 2012.

United States Senate. *Embassies as Command Posts in the Anti-Terror Campaign: A Report to the Committee on Foreign Relations*. United States Senate, Richard Lugar, Chairman, 109th Congress, Washington DC: US Government Printing Office, 15 December 2006.

United States Senate. Committee on International Relations and Committee on Foreign Relations. *Legislation on Foreign Relations Through 2002.* Current Legislation and Executive Orders, Washington DC: US Government Printing Office, 2003. http://transition.usaid.gov/ads/faa.pdf (accessed on 12 November 2012).

United States Congress. Public Law 99-433-1 October 1986, The Goldwater-Nichols Department of Defense Reorganization Act of 1986.

United States Government Accounting Office (GAO). "Plan Colombia: Drug Reduction Goals Were Not Fully Met, but Security Has Improved; US Agencies Need More Detailed Plans for Reducing Assistance." Washington DC: Government Printing Office, October 2008, GAO-09-07.

Uribe Velez, Alvaro. "An Interview with Alvaro Uribe Velez." *Prism* (National Defense University Press) 3, no. 3 (June 2012): 140-145.

Vaughn, Bruce. *Indonesia: Domestic Politics, Strategic Dynamics, and US Interests.* Washington DC: Congressional Research Service, 27 October 2010.

Wilson, Greg. "SOF Innovations Conference - Operating in the Gap." Presentation, Department of Defense Analysis, Naval Postgraduate Institute, Monterey, CA, 2011.

Doctrine

Department of Defense. Department of Defense Directive (DoDD) 3000.5, 28 NOV 2005, *USD (P), Military Support for Stability, Security, Transition, and Reconstruction (SSTR) Operations.* Washington DC: Government Printing Office, 2005.

———. Department of Defense Directive (DoDD) 5132.03,

DOD Policy and Responsibilities Relating to Security Cooperation. Washington DC: Government Printing Office, 24 October 2008.

Chairman, Joint Chiefs of Staff. Joint Publication 1-02, *Department of Defense Dictionary of Military and Associated Terms.* Washington DC: US Government Printing Office, 2012.

─────── . Joint Publication 5-0, *Joint Operational Planning.* Washington, DC: Government Printing Office, 2011.

Headquarters, Department of the Army. Army Doctrinal Publication 3-0, *Unified Land Operation.* Washington, DC: Government Printing Office, October 2011.

─────── . Army Doctrinal Publication 3-05, *Special Operations.* Washington, DC: Government Printing Office, 2012.

─────── . TRADOC PAM 525-3-0, *The Army Capstone Concept, 19 December 2012.* Washington DC: Government Printing Office, 2012.

─────── . Field Manual 5-0, *The Operations Process.* Washington DC: Government Printing Office, 2010.

Secondary Sources

Abuza, Zachary. *Militant Islam in Southeast Asia: Crucible of Terror.* Boulder, CO: Lynne Rienner Publishers, 2003.

Army, Department of the. *Army Activities in Underdeveloped Areas Short of Declared War.* Memorandum For the Secretary of the Army, Washington D.C.: Brigadier General Richard G. Stilwell, 13 October 1961.

Arquilla, John. "Guerilla Lit 101: Ten Books that are better than the Art of War." *Foreign Policy*, 24 September 2012. http://www.foreignpolicy.com/articles/2012/09/24/guerilla_lit_10 (accessed 26 September 2012).

_____. *Insurgents, Raiders and Bandits: How Masters of Irregular Warfare Have Shaped Our World.* Plymouth, UK: Ivan R. Dee, 2011.

Bassford, Christopher. "Clausewitz and His Works." In *The Reception of Clausewitz in Britain and America*, by Christopher Bassford, http://www.clausewitz.com/readings/Bassford/Cworks/Works.htm (accessed on 13 January 2013). New York: Oxford University Press, 1994.

Baumann, Andrea Barbara. "Clash of Organizational Cultures? The Challenge of Integrating Civil and Military Efforts in Stabilisation Operations." *RUSI Journal* 153, no. 6 (December 2013): 70-73. http://www.rusi.org/downloads/assets/Baumann.pdf (accessed 14 January 2013).

Berridge, Geoff, interview by Georgie Day. "Geoff Berridge on why we need diplomats." *The Browser: Writing worth reading*. (unpublished date): http://thebrowser.com/interviews/ geoff-berridge-on-why-we-need-diplomats (accessed 12 December 2012).

Birtle, Andrew J. *U.S. Army Counterinsurgency and Contingency Operations Doctrine 1942-1976.* Washington DC: Center of Military History, 2006.

Briscoe, Charles H., and Richard L. Kiper. *Weapon of Choice: U.S. Army Special Operations Forces in Afghanistan.* Fort Leavenworth, KS: Combat Studies Institute, 2003.

Briscoe, Charles H., Kenneth Y. Finlayson, Robert W. Jones, Jr., Cheryl A. Walley, Dwayne A. Aaron, Michael R. Mullins, and James Schroder. *All Roads Lead to Baghdad: Army Special Forces in Iraq.* Boulder, CO: Paladin Press, 2007.

Brooks, Rosa. "Obama Needs a Grand Strategy." *www.foreignpolicy.com* (23 January 2012). http://www.foreignpolicy.com/articles/2012/23/obama_needs_a_grand_strategy (accessed 16 April 2012).

Butler, Dave. "Lights Out: ARSOF Reflect on 8 Years in Iraq." *Special Warfare* (The John F. Kennedy Special Warfare Center and School) 25, no. 1 (January-March 2012): 28-34.

CBS Evening News. "Colombian to Aid US in Taliban Fight." *CBS Evening News online*, 27 July 2009. http://www.cbsnews.com/8301-18563_162-5192173.html (accessed 28 January 2013).

Chandrasekaran, Rajiv. *Little America: The War Within the War for Afghanistan*. New York: Random House, Inc., 2012.

Clausewitz, Carl von. *On War*. 1976. The original German text from 1832 was translated, edited and published as *On War* by Michael Howard and Peter Paret. Princeton: Princeton University Press 1976.

Cole, Ronald H., Walter S. Poole, James F. Schnabel, Robert J. Watson, Willard J. Webb. *The History of the Unified Command Plan*. Washington DC: Joint History Office, Office of the Chairman of the Joint Chiefs of Staff, 1995.

Comer, Charles "Ken". "Leahy in Indonesia: Damned if you do (and even if you don't)." *Asian Affairs: An American Review* (Francis & Taylor Group, LLC) 37 (2010).

Creveld, Martin van. "Napoleon and the Dawn of Operational Warfare." In *The Evolution of Operational Art*, by John Andreas and van Creveld, Martin Olsen, 9-34. New York: Oxford University Press, 2011.

Defense Institute of Security Assistance Management (DISAM). *The Management of Security Cooperation (Greenbook)*. 31st edition. Washington, DC, February 2012.

Dolman, Everett Carl. *Pure Strategy: Power and Principle in the Space and Information Age*. New York: Frank Cass, 2005.

Dozier, Kimberly. "US commandos boost numbers to train Mexican forces." *NBC News*, 17 January 2013. http://www.msnbc.msn.com/id/50496049#.UQQ1v6ViZV

(accessed 26 January 2013).

Echevarria II, Antulio. "War's Second Grammar." *Strategic Studies Institute*, October 2009.

Erwin, Sandra I. "Special Operations Command Seeks a Bigger Role in Conflict Prevention." *National Defense Magazine* (29 November 2012). http://nationaldefensemagazine.org/blog/Lists/Posts/Post.aspx?ID=983> (accessed 17 December 2012).

Faint, Charles and Michael Harris. "F3EAD: Ops/Intel Fusion 'Feeds' the SOF Targeting Process." *Small Wars Journal* (January 2012). http://smallwarsjournal.com/jrnl/art/f3ead-opsintel-fusion-"feeds"-the-sof-targeting-process.

Farrelly, Nicholas. "Counting Thailand's Coups." *New Mandala*, (8 March 2011). http://asiapacific.anu.edu.au/newmandala/2011/03/08/counting-thailands-coups/ (accessed 21 December 2012.)

Finlayson, Kenneth. "Colombian Special Operations Forces." *Veritas: Journal of Army Special Operations History* 2, no. 4 (November 2007): 56-59.

Forero, Juan. "In Colombian Jungle Ruse, US Played a Quiet Role; Ambassador Spotlights Years of Aid, Training." *Washington Post*, 9 July 2008, http://www.washingtonpost.com/wp-dyn/content/story/2008/07/08/ST2008070803342.html?pos= (accessed 17 January 2013).

Fraser, General Douglas M. "Posture Statement of General Douglas M. Fraser." United States Air Force, Commander, United States Southern Command, before the 112th Congress House Armed Services Committee, 6 March 2012.

Galula, David. *Counterinsurgency Warfare: Theory and Practice*. Saint Petersburg, Florida: Hailer Publishing, 2005. First published in 1964.

Geertz, Clifford. *The Interpretation of Cultures.* New York, New York: Basic Books, 1973.

Gharajedaghi, Jamshid. *Systems Thinking: Managing Chaos and Complexity: A Platform for Designing Business Architecture.* Burlington, MA: Elsevier, 2006.

Gleiman, Kenneth LTC. "Operational Art and the Clash of Organizational Cultures: Postmortem on Special Operations as a Seventh Warfighting Function." Monograph, School of Advanced Military Studies, United States Army Command and General Staff College, 2011.

Gonsalves, Marc, Keith Stansell, Tom Howes, and Gary Brozek. *Out of Captivity: 1,967 Day in the Colombian Jungle.* New York: Harper Collins Publishers, 2009.

Gray, Colin S. *Another Bloody Century: Future Warfare.* London, UK: Phoenix, 2005.

⎯⎯⎯⎯. *Modern Strategy.* Oxford, UK: Oxford University Press, 1999.

⎯⎯⎯⎯. *Strategy for Chaos: Revolutions in Military Affairs and the Evidence of History.* Portland: Frank Cass Publishers, 2002.

⎯⎯⎯⎯. War, Peace and International Relations. New York: Routledge, 2007.

Guevara, Che. *Guerilla Warfare.* Lincoln, NE: University of Nebraska Press, 1998.

Herspring, Dale R. *The Pentagon and the Presidency: Civil-Military Relations from FDR to George W. Bush.* Lawrence, KS: University Press of Kansas, 2005.

Holloway III, Admiral James L. *Iran Hostage Rescue Mission Report (Unclassified).* Special Operations Review Group, Washington. DC: The Navy Department Library, 23 August 1980. http://www.history.navy.mil/library/online/hollowayrpt.htm (accessed 14 January 2013).

Jane's online. "Fuerzas Armadas Revolucionares de Colombia (FARC), Ejercito de Liberacion Nacional, and Autodefensas Unidas de Colombia." www.janes.com (accessed 17 January 2013).

Joint Special Operations University. *Special Operations Reference Manual*. 3rd Edition. MacDill Air Force Base, FL: Joint Special Operations University Press, 2011.

Jomini, Antoine Henri. *The Art of War*. Translated by J. B. Lippincott & Co. London: Lionel Leventhal Limited, 1992.

Jones Jr., Robert W. "Plan Colombia and Plan Patriota: The Evolution of Colombia's National Strategy." *Veritas: Journal of Army Special Operations History* 2, no. 4 (November 2007): 60-64.

Jones, Seth. "Going Local: The Key to Afghanistan." *The Wall Street Journal*, 7 August 2009.

Kamen, Al. "The End of the Global War on Terror." *The Washington Post*, 24 March 2009: http://voices.washingtonpost.com/44/2009/03/23/the_end_of_the_global_war_on_t.html (accessed 14 January 2013).

Keating, Timothy. "US Forward Presence in the Asia-Pacific Region." Audio podcast from CSIS panel discussion (iTunes, accessed 28 January 2012, 1:10:00 to 1:21:00). Center for Strategic and International Studies (CSIS), 24 September 2012.

Kennedy, John F. "JFK on Special Forces." *Address to the United States Military Academy graduating class*. West Point, NY: You Tube, 6 June 1962. http://www.youtube.com/watch?v=7FVrpiG7haE (accessed 12 November 2012).

Kenney, Michael. From *Pablo to Osama: Trafficking and Terrorist Networks, Government Bureaucracies, and Competitive Adaptation*. University Park, PA: The Pennsylvania State University Press, 2007.

Kiras, James D. *Special Operations and Strategy: From World War II to the War on Terrorism*. New York, NY: Routledge, 2006.

Kirila, Robert. "Ahead of the Guns: SOF in Central America." *Special Warfare* (US John F. Kennedy Special Warfare Center and School) 25, no. 4 (October-December 2012): 15-18.

Kissinger, Henry. *Diplomacy*. New York: Simon & Schuster, 1994.

Krause, Micheal D., and R. Cody Phillips. *Historical Perspectives of the Operational Art*. Washington, DC: Center for Military History, 2005.

Kugler, Richard L. New Directions in National Security Strategy, Defense Plans, and Diplomacy: A Review of Strategic Documents. Washington, DC: National Defense University Press, 2011.

Lawrence, T.E. *Seven Pillars of Wisdom*. London: Penguin Group, 1962. First published (privately) in 1926.

Lawson, Bryan. *How Designers Think: The Design Process Demystified*. Oxford: Elsevier Ltd, 2006.

The Good, The Bad and The Ugly. Motion Picture. Directed by Sergio Leone. Produced by Metro Goldwyn Meyer. 1966.

Lujan, Fernando M., Major, US Army. "Light Footprints: The Future of American Military Intervention," Center for New American Studies (CNAS), Voices from the Field, March 2013.

Lumbaca, J. "Lumpy." "The US-Thailand ARSOF Relationship." *Special Warfare* (U.S. John F. Kennedy Special Warfare Center and School) 25, no. 1 (January-March 2012): 52-53.

Luttwak, Edward. *The Logic of War and Peace*. Cambridge, MA: Belknap Press, 2003.

Manwaring, Max G. "The Strategic Logic of the Contemporary

Security Environment." Monograph, U.S. Army War College, Carlisle Barracks, PA: Strategic Studies Institute, December 2011.

———. "Ambassdor Stephen Krasner's Orienting Principle for Foreign Policy (and Military Management) - Responsible Sovereignty." Monograph, U.S. Army War College, Carlisle Barracks, PA: Strategic Studies Institute, April 2012.

Marks, Thomas. "A Model Counterinsurgency: Uribe's Colombia (2002-2006) vs. FARC." *Military Review*, March-April 2007: http://usacac.army.mil/cac2/call/docs/11-15/ch_6.asp (accessed 17 January 2013).

———. "Colombia: Learning Institutions Enable Integrated Responses." *Prism* (National Defense University Center for Complex Operations) 1, no. 4 (September 2010): 127-146.

McChrystal, Stanley. "Becoming the Enemy: To win in Afghanistan, we need to fight more like the enemy." *Foreign Policy*, March/April 2011: 66-70.

McDougall, Walter. *Promised Land, Crusader State*. New York: Houghton Mifflin Company, 1997.

Menning, Bruce W. "Operational Art's Origins." In *Historical Perspectives of the Operational Art*, by Michael D. and Phillips, R. Cody Krause, 4-18. Washington DC: Center for Military History, 2005.

Metz, Steven. "Strategic Horizons: U.S. Army Prepares for the Human Domain of War." *World Politics Review* (November 2012). http://www.worldpoliticsreview.com/articles/12481/strategic-horizons-u-s-army-prepares-for-human-domain-of-war (accessed 13 January 2013).

Millett, Allan R. and Peter Maslowski. *For the Common Defense: A Military History of the United States of America*. New York: The Free Press, 1984.

Mintzberg, Henry. *The Rise and Fall of Strategic Planning.* New York: The Free Press, 1994.

Mintzberg, Henry and James Brian Quinn. *Readings in the Strategy Process.* 3rd Edition. Upper Saddle River, NJ: Prentice Hall, Inc., 1998.

Nye, Joseph. *The Future of Power.* New York: Public Affairs, 2011.

Olsen, John Andreas, and Martin Van Creveld. *Evolution of the Operational Art: From Napoleon to Present.* Oxford, United Kingdom: Oxford University Press, 2011.

Paddock, Alfred H. *US Army Special Warfare: Its Origins.* Lawrence, KS: University Press of Kansas, 2002.

Parnass, Sarah. "Hillary Clinton Endures Brusque Questioning at Hearings." *ABC News online,* (23 January 2013). http://abcnews.go.com/Politics/OTUS/hillary-clinton-endures-brusque-questioning-hearings/story?id=18292329 (accessed 29 January 2013).

Payne, Robert. *The Marshall Story: A Biography of General George C. Marshall.* New York: Prentice Hall, 1951.

Perez Jr., Celestino, ed. *Addressing the Fog of CoG: Perspectives on the Center of Gravity in US Military Doctrine.* Combat Studies Institute, US Army Command and General Staff College, Fort Leavenworth, KS: Combat Studies Institute Press, 2012.

Project for National Security Reform. "Ensuring Security in an Unpredictable World: The Urgent Need for National Security Reform." *Project for National Security Reform website* (July 2008). www.pnsr.org (accessed 15 January 2013).

Ramsey III, Robert D. *From El Billar to Operations Fenix and Jaque.* Occasional Paper 34, Combat Studies Institute, Fort Leavenworth, KS: Combat Studies Institute Press, 2009.

Rempe, Dennis M. "The Past as Prologue: A History of the

US Counter-Insurgency Policy in Colombia 1958-1966." Monologue, Strategic Studies Institute, March 2002.

Renehan, Edward J. *The Monroe Doctrine: The Cornerstone of American Foreign Policy*. New York: Oxford University Press, 2007.

Reuters. "CIA Chief: Yemen Qaeda most dangerous." *Reuters online news service* (13 September 2011). http://www.reuters.com/article/2011/09/13/us-usa-security-qaeda-idUSTRE78C3G720110913 (accessed 15 January 2013).

Ricks, Tom. *Fiasco: The American Military Adventure in Iraq*. New York: Penguin Books, 2006.

Roth, Brigadier General Gunter R. "Operational Thought from Schlieffen to Manstein." In *Historical Perspectives of the Operational Art*, by Michael D. Phillips and R. Cody Krause. Washington DC: Center for Military History, 2005, 149-166.

Sacquety, Troy J. "Colombia's Troubled Past." *Veritas: Journal of Army Special Operations History* 2, no. 4 (November 2007): 8-14.

Saum-Manning, Lisa. "VSO-ALP: Comparing Past and Current Challenges to Afghan Local Defense." *Small Wars Journal* (December 2012). http://smallwarsjournal.com/jrnl/art/vsoalp (accessed on 3 January 2012).

Schmitt, Eric. "Elite Military Forces are Denied in Bid for Expansion." *The New York Times*, 4 June 2012. http://www.nytimes.com/2012/06/05/world/special-ops-leader-seeks-new-authority-and-is-denied.html?hp&_r=0 (accessed 15 December 2012).

Scott, James. "Interbranch Rivalry and the Reagan Doctrine in Nicaragua." *Political Science Quarterly* 112, no. 2 (1997): 237-260.

Sharp, Gene. *From Dictatorship to Democracy: A Conceptual*

Framework for Liberation. New York: The New Press, 2002. Originally published in 1993.

Showalter, Dennis. "Prussian-German Operational Art, 1740-1943." In *The Evolution of Operational Art*, by Michael D. Phillips and R. Cody Krause. New York: Oxford University Press, 2011, 35-63.

Simons, Geoff. *Colombia: A Brutal History.* London: Saqi Books, 2004.

Smith, Rupert. "Epilogue." In *The Evolution of Operational Art*, by John Andreas and van Creveld, Martin Olsen, 236. New York: Oxford University Press, 2011.

———. The Utiilty of Force: The Art of War in the Modern World. New York: Vintage Books, 2005.

Svechin, Alexsandr. *Strategiia (Strategy).* Translated by Kent D. Lee. Minneapolis: Eastview Publications, 1992.

The Nation. "US cuts off millions in military aid to Thailand." 29 September 2006: http://nationalmultimedia.com/2006/09/29/headlines/headlines_30014950.php (accessed 21 December 2012).

Taber, Robert. *War of the Flea: The Classic Study of Guerilla Warfare.* Washington DC: Brassey's Inc., 2002. Originally published in 1965.

Trinquier, Roger. *Modern Warfare: A French View of Counterinsurgency.* London: Pall Mall Press Ltd., 1964. First published in France in 1961 as *La Guerre Moderne.*

Tse-Tung, Mao. *On Guerilla Warfare.* United States: Praetorian Press LLC, 2011. Originally published in Chinese in 1937.

Tzu, Sun. *The Art of War*, translated by Ralph D. Sawyer (Boulder, Colorado: Westview Press, Inc., 1994). Original Chinese text dated to 500 B.C. era.

US Department of State. www.state.gov.

Wald, Charles F. "New Thinking at USEUCOM: The Phase

Zero Campaign." *Joint Forces Quarterly*, 4th quarter 2006: 72-75.

Waller, Douglas. *Wild Bill Donovan: The Spymaster who created the OSS and modern American espionage.* New York: Simon & Schuster, Inc., 2011.

Whitlock, Craig. "U.S. counterterrorism efforts in Africa defined by a decade of missteps." *Washington Post*, February 4, 2013. http://www.washingtonpost.com/craig-whitlock/2011/02/28/AB5dpFP_page.html (accessed 9 February 2013).

CPSIA information can be obtained
at www.ICGtesting.com
Printed in the USA
FFOW02n0610090116
20204FF